Social Responses to Large Technical Systems
Control or Anticipation

NATO ASI Series

Advanced Science Institutes Series

A Series presenting the results of activities sponsored by the NATO Science Committee, which aims at the dissemination of advanced scientific and technological knowledge, with a view to strengthening links between scientific communities.

The Series is published by an international board of publishers in conjunction with the NATO Scientific Affairs Division

A Life Sciences **B Physics**	Plenum Publishing Corporation London and New York
C Mathematical and Physical Sciences **D Behavioural and Social Sciences** **E Applied Sciences**	Kluwer Academic Publishers Dordrecht, Boston and London
F Computer and Systems Sciences **G Ecological Sciences** **H Cell Biology** **I Global Environmental Change**	Springer-Verlag Berlin, Heidelberg, New York, London, Paris and Tokyo

Series D: Behavioural and Social Sciences - Vol. 58

Social Responses to Large Technical Systems

Control or Anticipation

edited by

Todd R. La Porte

Department of Political Science and
Institute of Governmental Studies,
University of California,
Berkeley, U.S.A.

Kluwer Academic Publishers

Dordrecht / Boston / London

Published in cooperation with NATO Scientific Affairs Division

Proceedings of the NATO Advanced Research Workshop on
Social Responses to Large Technical Systems:
Regulation, Management, or Anticipation
Berkeley, California, U.S.A.
October 17–21, 1989

ISBN 0–7923–1192–2

Published by Kluwer Academic Publishers,
P.O. Box 17, 3300 AA Dordrecht, The Netherlands.

Kluwer Academic Publishers incorporates the publishing programmes of
D. Reidel, Martinus Nijhoff, Dr W. Junk and MTP Press.

Sold and distributed in the U.S.A. and Canada
by Kluwer Academic Publishers,
101 Philip Drive, Norwell, MA 02061, U.S.A.

In all other countries, sold and distributed
by Kluwer Academic Publishers Group,
P.O. Box 322, 3300 AH Dordrecht, The Netherlands.

Printed on acid-free paper

Printed in the Netherlands

CONTENTS

PREFACE

This volume stems from the efforts of scholars who seek to understand the social dynamics of large technical systems. The purpose is to develop concepts and empirical knowledge concerning the dynamics of such systems, with particular emphasis on the processes of control and/or management in a variety of national settings, and to improve the basis of public policy so that future developments might be less distressing in consequence and more shaped to the desires of their "host" societies.

One vehicle for this enterprise is a series of international conferences on the Evolution and Dynamics of Large Technical Systems (LTSs). This series was instituted to encourage the coalescence of the multidisciplinary group of scholars who are actively engaging in the empirical study of these phenomena. Their disciplines span history, sociology, political science, and economics studies. They come Australia, France, the Netherlands, Norway, Sweden, the United Kingdom, the United States, and West Germany. And they possess strong backgrounds in the empirical study of specific technical areas and a taste for conceptual and theoretical integration.

The first conference, "The Development of Large Technical Systems—Theoretical Approaches, Empirical Cases, and International Comparison," was sponsored by and held at the Max Planck Institute for Social Research, Cologne, West Germany, in November 1987. Its papers appear in *The Development of Large Technical Systems*, edited by Renate Mayntz and Thomas P. Hughes. The present volume includes revisions of papers presented at the the second conference, "Societal Responses to the Development of Large Technical Systems: Control, Adjustment or Adaptation," held 17–21 October 1989 at the University of California at Berkeley, U.S.A. (The conference agenda and a list of participants appear in the Appendix).

The Berkeley conference was intense, yeasty, often dramatic, and a bit harrowing, with considerable warmth of feeling among the participants—a necessary glue in holding together a group of disparate perspectives. Our welcoming dinner began about twenty minutes after a major earthquake struck nearby. As the dinner progressed on the more or less unaffected university campus, news slowly filtered in about the full effects of the quake. Several of our number were absent. Two were unaccounted for; later, it turned out that they narrowly escaped the collapse of a major freeway as they were returning their auto to the airport. Conferees, staying in a slightly damaged hotel nearby, faced the next day without hot water. Two U.S. participants could not join us the next day due to disrupted air transport.

The earthquake and the startling effectiveness of public agencies' emergency responses and remarkable citizen outpouring of aid gave an extraordinary cast to our discussions. One session the next day was spent comparing videotapes of the media response immediately after the quake with reporting a day later. It was an occasion to reflect on comparative perceptions of events and organizational responses. Overall the discussions were vigorous, lively, and fruitful, especially given the circumstances.

A conference of this scale needs a good deal of support and skillful assistance to facilitate keen interchange. Gratefully, we acknowledge the financial support of the NATO Advanced Research Workshop Program. Craig Sinclair, the program's director, was encouraging from the outset. The NATO program provided the resources for foreign travel. The U.S. Congress Office of Technology Assessment, through the good offices of John Andelin, its assistant director, supported the travel of U.S. participants. The Institute of International Studies, University of California, Berkeley, provided admirable local support. They gave superb assistance in administrative matters, local hospitality, and editorial help in getting the papers in shape for publication. Karin Beros, Katherine Merrill, Nadine Zelinsky, and Stephen Pitcher were particularly effective.

Todd R. La Porte
Berkeley, California
September 1990

Foreword

THE CHALLENGE OF UNDERSTANDING LARGE TECHNICAL SYSTEMS

Todd R. La Porte
Department of Political Science and
Institute of Governmental Studies
University of California at Berkeley

The number, scale, complexity, and range of large technical systems (LTSs) has increased steadily over the past hundred years. Their importance for social development and public policy matters is unquestioned. The most significant properties of such systems are their growing scale, increasingly intensive knowledge requirements, tightening patterns of functional interdependence within major productive or service segments, and expanding networks of cooperation and control. Yet these phenomena are not well understood. They confound engineers, social scientists, historians, economists, policy planners, and political leaders. Social and organizational theories falter in the face of complex, interdependent relationships. Nor do historical or economics-based models of technological change give policymakers a firm theoretical basis for their decisions.

Governments make technological policy with little help from historical experience or systematic social or economic theory. Policymakers need to know more about the ways in which complex technologies evolve, and the effectiveness of various policy instruments in shaping that process. The characteristics and behavior of technical systems may differ over time, or display similar patterns over their life histories, such as those of the telephone or electrical power systems, ballistic missiles, air traffic control, and computer systems. Many of the policy effects of technological systems appear well after the first years of deployment. More certain knowledge of both the evolution and dynamics of such systems would make policy more effective and less fraught with surprise. Likewise, legislators should expect to craft different types of regulatory instruments when dealing with technological systems in different stages of innovation and development. Such instruments are strongly shaped by their particular political and institutional histories and appear to vary significantly among advanced and less developed nations.

Conceptual inadequacy and a meager sense of history would be mainly a matter of academic interest and "politics as usual"—if the benefits of such systems were modest, and the consequences of their failures limited. But in a growing number of areas neither of these conditions hold: benefits are not piecemeal, nor breakdowns mainly instructive. Rather, benefits of many large-scale systems are themselves large-scale, dispersed, and often quite generous. And significant failures (or even the prospect of them) evoke widespread concern, sometimes great fear. Indeed, for an increasing number of LTSs, benefits are conditioned on continuously reliable operations, and, in some cases—such as nuclear power—nearly failure-free operations. These requirements pose an extraordi-

T. R. La Porte (ed.), Social Responses to Large Technical Systems, 1–4.

nary situation; they exert very strong demands for remarkable performance at a time when there is scant understanding of the conditions relevant to such operations. Yet we are becoming increasingly dependent on systems with just such characteristics.

Large technical systems excite—and confound—all advanced industrial societies. The policy challenges they provide are a regular concern of all nations striving to take on the mantle of advanced industrialism. Yet there is scant systematic knowledge about their patterns of evolution or internal dynamics, and limited systematic knowledge about their effects upon different political systems. Building on earlier primarily historically oriented work,[1] the authors in this volume turned to the problem of controlling, managing, or adjusting large technical systems during the deployment process.

When Societies Respond to the Deployment of Large Technical Systems[2]

When technical systems promise substantial benefits, they frequently are developed to large scale. Some LTSs (e.g., the air transport industry) produce enormous benefits as they approach mid-stages of deployment and continue to do so as they reach full market maturity. A few LTSs develop in the midst of controversy and are still in question as they become mature, e.g., the U.S. nuclear industry. Other LTSs (e.g., the chemical industry) produce substantial benefits in the early and mid-stages of deployment only to become seen as the source of considerable distress as they grow to very large, mature scale and produce substantial environmental and social disruptions.

Nation-states struggle simultaneously to maintain technologically based benefits and to redirect the character of LTSs as they advance toward fully deployed systems in order to mitigate surprising, negative consequences mature technical systems can have. How have the leading institutions of advanced industrial societies responded to the successes, surprises, and negative effects of technical systems as they have flowered and grown toward large scale? These responses vary significantly from one advanced industrial country to another. What explains these differences? Do different technical systems provoke different institutional responses? To what degree have such responses produced the hoped-for results?

Our interest is in the ways societies attempt to manage or shape deployed LTSs; that is, those LTSs that have reached mid or mature stages of deployment and are discovered either to have substantially greater benefits or negative social, economic, or environmental consequences than were expected.[3] Societies often respond to unusual benefits with unrestrained enthusiasm and pell-mell, unregulated deployment to full scale. Negative surprises raise different problems. Responses range from "after the fact" mitigation of environmental damage, to attempts to anticipate and "design out" problems "before the fact." In more detail, societies respond variously by attempting to:

- Prevent any regulation or deregulate a technical area if there had been regulations;
- Provide governmental subsidies and legal protection for the deployment of a particularly attractive technical system, (e.g., military and governmental

procurement processes);

- Develop an analytical capacity, (e.g., policy analysis or technology assessment), to forecast or anticipate potential effects of a LTS during and after the deployment process so as to "design away" effects that are not politically desirable;
- Legislate regulations directed toward moderating the behavior of deployers in the early deployment stages (e.g., environmental impact analysis process, and/or demands for failure-free technical and organizational performance), so that the potentials for subsequent undesirable effects is diminished;
- Enforce punitive economic and legal regulations after the damage or surprising consequence of an LTS becomes evident (e.g., liability litigation processes).

These responses lead to particular knowledge requirements and institutional imperatives. They are likely to vary as a function of the properties of the technical system and/or the society within which it is being deployed. Conference papers ventilate the intellectual and policy issues involved in some of these responses. The intent is to develop more systematic understanding of their dynamics and the limits of their effectiveness. Approaches made use of both history and contemporary social science, and the comparison of technologies both within and among a variety of advanced industrial societies. Particular emphases included the problems of system failures and their external effects, the side-effects of full deployment, momentum/autonomy/uncontrolled developments and liability, types of system controls (e.g., bureaucratic and computer-aided controls, interactions with organization cultures, and regulatory legislation.

Finally, there is an emphasis on the need for: 1) increasing precision in specifying the phenomenon—technical systems—in social scientific, as well as engineering, terms; 2) examining models of technological and political change or evolution in order to explicate concepts used in them, including models of how political decisions play a role in the development of technological systems and the uses of case histories to define the models and concepts; 3) explaining similarities and/or differences in the dynamics and social properties of apparently quite different technical systems developed in different political settings during various historical periods; and 4) exploring the reciprocal contributions of history and social science, especially the conceptual and methodological conditions conducive to effective contributions of historians and social scientists to the understanding of technological systems.

NOTES

1. See *The Development of Large Technical Systems*, ed. R. Mayntz and T. P. Hughes (Frankfurt: Campus Verlag, and Boulder: Westview Press, 1988). Cf. *The Social Construction of Technological Systems: New Directions in the Sociology and History of Technology*, ed. W. E. Bijker, T.

P. Hughes, and T. Pinch (Cambridge, Mass.: MIT Press, 1987).

2. From the conference call for papers, January 1988.

3. This contrasts to interests in the full train of steps from invention of technical possibilities to the fully matured large scale producing benefits and/or distress. These steps include 1) invention, 2) feasibility, 3) initial start-up, 4) mid-stage deployment, 5) mature technical systems at scale, and 6) possibly the discovery of unanticipated effects.

REFERENCES

Bijker, W. E.; Hughes, T. P.; and Pinch, T., eds. 1987. *The Social Construction of Technological Systems: New Directions in the Sociology and History of Technology.* Cambridge, Mass.: MIT Press.

Hughes, Thomas P. 1983. *Networks of Power: Electrification in Western Society, 1880–1930.* Johns Hopkins University Press.

Mayntz, R., and Hughes, T. P., eds. 1988. *The Development of Large Technical Systems.* Frankfurt: Campus Verlag; and Boulder: Westview Press.

LARGE TECHNICAL SYSTEMS, REAL-LIFE EXPERIMENTS, AND THE LEGITIMATION TRAP OF TECHNOLOGY ASSESSMENT: THE CONTRIBUTION OF SCIENCE AND TECHNOLOGY TO CONSTITUTING RISK PERCEPTION

Peter Weingart
Center for Interdisciplinary Studies
University of Bielefeld

Definitions of Risk and Structures of the Discourse on Risk

For a long time engineers and statisticians insisted that risk calculations based on the frequency rate of accidents and the volume of damage that actually occurred were rational, and castigated the public for their irrational fears. This alleged rationality has since been exposed as uninformed in sociological and psychological terms as well as politically naive. We now know that social awareness and acceptance of risks and dangers are not related to the question of whether the latter will actually appear. The implicitly or explicitly proposed "expertocratic" model of enlightenment is based on the assumption that if only the public had an understanding of the scientific-technological facts, they would necessarily arrive at the same conclusions as the engineers and scientists. This view falters due, in part, to the public's manifest "disinclination" to understand. In addition, the psychologists have given the blessing of perceptual psychology to the difference between the "hard facts" (as presented by the scientists and engineers) and the individual perceptions of risk (on the part of the non-experts). This has knocked the bottom out of any delegitimation attempted by the scientific-technological monopoly on rationality.[1] Sociologists have even gone one step further to explain the validity of several incompatible rationalities. Wynne talks of a "dual society" that owes its origin mainly to the rationalist myth of science and could be likened to the situation of peasants and development experts in the Third World: the greatest risks are of a social nature and derive from the fact that control has to be placed in the hands of "experts" whose values and objectives they do not understand and whom they therefore do not trust (Wynne 1982: 137). The division is between the rationality claimed by societal institutions for their quantifiable goals and the rationality of risk avoidance claimed by the "non-experts."

Taking this further, Perrow (1984: 315f.) has identified the "absolute rationality" of economists and engineers and the "limited" rationality of a growing number of chastened risk researchers as well as a "social" or "cultural" rationality that guides everyday action.[2] In their culture/sociological approach, Douglas and Wildavsky (1982) have undertaken a more sophisticated categorization and assigned types of rationalities to types of societal institutions or groups which explain in sociological terms the differences in perception and thus document their "justification." This approach makes possible an institutional identification even of those sociological analyses that either seek to relate the risk debate to the "factual" risks (as the fashionably radical critique of science would have it,

T. R. La Porte (ed.), Social Responses to Large Technical Systems, 5–17.

"objective counter-force of danger"; Beck 1988: 636), or to dismiss it as an expression of diffuse anxieties befitting the sophisticated irony of constructivism typical of Luhmann (Luhmann 1986: 237ff).

The multiplication of rationalities has now reached science itself in the form of an epistemological paradigm shift to a "radical constructivism" which, through its de-trivialized misunderstanding, seems to suggest a surrender of the rationality monopoly by science itself. The existence of a diversity of rationalities corresponds to a multiplication of equally legitimate viewpoints, implying a leveling of epistemological and methodological privileges. To what extent this process is transcending the boundaries of science (especially the social sciences) is a moot point. That there have been some consequences is obvious when we look at the connections between this development and the social movements in the wake of the "New Age."

Following the logic of systems theory, or by just considering the evidence of multiple rationalities, one could conclude that just as the perception of risks is dependent on the different frames of reference of social systems, so their causation and the assignment of potential consequences, too, is a question of an observer's vantage point. In a very general, trivial sense of the term, this is of course true. The risks featured in the politically fired debate are more specific, however: these risks are "caused" by new technologies and/or they are identified as such by a science, i.e., risk assessment. The scientific system has a special role in this context since it produces scientific "truths" that serve as "substance" and orientation of individual perception, in the formation of a collective opinion and in all kinds of policy choices—however mistaken such truths may be. In identifying the reasons for the dynamic of the risk debate it seems opportune, therefore, to look at the scientific system itself. The special links between science, politics, and economics, that combine to give risks their "gestalt," will not be discussed here except insofar as they relate to the topic of this paper.

I will proceed by arguing—referring in part to recent works on risk assessment—that, regardless of the distinction between institutionally assignable types of risk perception and types of rationalities, the dividing line that is relevant in terms of legitimation is one that runs between "risk-producing" institutions and a "risk-affected" public. The scientific system comes under pressure of legitimation to the extent that it is seen as the originator of risky technologies. This pressure will increase because conditions inherent in the scientific system will lead to its progressive "technicalization," i.e., a systematic production of technological risks. In the process, the scientific system will inevitably be drawn into a political maelstrom since the risks themselves will become the object of conflict over technological conceptions and the diffusion of consequences among social institutions, on one side, and social (protest) movements, on the other. This is paradoxical because the protagonists on both sides nevertheless take recourse to science in trying to define the risks in question. Unfortunately, the positive effect of an emerging consensus, as anticipated by science and risk assessment, in particular, is not going to materialize. This is the case not only because of the impact of politicization, but also because the identification of risks has already been claimed by the self-referential and seemingly unrestrained dynamic of science.

Risks vs. Dangers: Someone's Risks Pose Dangers for Someone Else

The serious concern for a "social rationality" and the assignment of different rationalities to social groups can be accounted for more precisely in theoretical terms. Luhmann has laid open the historical roots underlying the introduction of the semantic distinction between dangers and risks. This defines as a danger "any not too improbable, negative effect on one's own life," and as a risk any disadvantage that must be attributed to an individual's own decision. His thesis should be seen in relation to constructivist epistemology according to which science and technology are preoccupied with testing their own expectations of nature, thus turning our interaction with nature into a constructive relationship. In other words, rather than thinking in terms of "dangers" in the sense of uncontrollable, unexpected natural events that occur independent of our actions, we have to think in terms of "risks" that are an outcome of our behavior, especially our production of knowledge and technology (Luhmann 1988: ch. 8).[3] Every additional scientific communication and every additional technological realization generates previously non-existent opportunities for decision-making, thus implying decisions characterized by uncertainty that involve not only choosing between different options, but also determining potential consequences. In this general sense, risks are constantly increasing because the knowledge process broadens the diversity of our "constructions." The transition from dangers to risks, therefore, applies in a general sense to the overall evolution of world society. In terms of concrete, institutionally reconstructable experience, it is primarily a phenomenon encountered in (modern industrialized) societies which have a differentiated scientific system.

The transformation of dangers into risks has become socially relevant and conceivable only through the emergence of a system of science sufficiently specialized to produce theoretical and empirically tested knowledge and to make these data available for the legitimation of political and economic decisions as well as for technological developments. Science and technology (as a transformation of scientific and empirical knowledge) and the concomitant decision-making opportunities are thus constantly producing "risks" that constitute "dangers" to other systems in the sense discussed earlier.

We are confronting a paradox inasmuch as the legitimation of science and technology rests on the assumption that the latter helps to reduce or eliminate dangers. Science and technology represent specialized institutions that are legitimated to transform "dangers" into "risks." It is obvious that this transformation role carries special obligations in terms of legitimation, especially concerning the rules and procedures guiding the operations of the scientific system, and the development and testing of new technologies. Whenever these rules and procedures are rendered inoperative, and/or whenever the "production" of risks shifts from specialized "institutions" to individual groups within or outside the scientific system whose riskful actions pose a danger for others, then the intra-scientific consensus, as reflected by specific rules and procedures, and the social consensus which is contingent upon it, are in danger of disintegrating. It follows that this would invariably undermine the legitimation of science and technology itself.

Two causes of this development will be looked at more closely—the conditions under which scientific rules and procedures are rendered ineffective, and those characteristics of technology that cause the risks produced by science and technology to be perceived as a danger to the rest of society.

The Erosion of the Institutional Basis of Scientific-Technological Legitimation

The laboratory is the physical correlate of the differentiation of science in terms of an experimental interaction of society with knowledge. It guarantees, so to speak, a public approval of this kind of riskful activity. Strictly speaking, any application of knowledge in the realization of a new technology produces more knowledge, turning any transfer of technology from the laboratory to the social domain into an undertaking fraught with uncertainties. While this procedure remains the "norm," the isolation of the laboratory as an experimental space that shields society from the consequences of experiments can be accepted as a sufficient guarantee of safety. The scientific laboratory stands for the safety of research and technological development and thus for their legitimation. Should the windows and doors of the laboratory develop leaks, however, this legitimation would be challenged. That this is not merely a metaphor is borne out by the discussions on the safety standards of genetic engineering at the time of the initial recombinant DNA experiments which focused explicitly on the "containment security" of the laboratories concerned.[4] As long as laboratories were considered "safe" and scientists and engineers, as nineteenth-century biographies would have it, served progress by "experiments performed on themselves," the risks involved only endangered the experimenters. The legitimation of high-risk research rested precisely on the fact that scientists endangered only themselves. At the same time, this gave rise to hopes that the knowledge gained along the way would eliminate future dangers.

Experiments to obtain new knowledge were conducted in the laboratory. New techniques and their functioning were tested under laboratory conditions. The effects of any, malfunctions were restricted to the lab, but, contrary to previous practice, the protected space of the laboratory is now increasingly being left behind with the result that "society is becoming a laboratory."[5] This movement in the society is part of the dynamic inherent in the scientific system and can boast a certain tradition. As early as the turn of the century, the physician Wilhelm Schallmayer, for example, developed a blueprint for a comprehensive hereditary record. By doing so he explicitly stated that this would both be a condition for confirming the validity of hereditary theories and for their practical implementation in the shape of eugenic policies. The shift of experiments, and thus risks, from the laboratory to the social sphere was occasioned by those technologies whose effects, either through malfunction, or even normal operation, may be modeled but cannot be tested in the laboratory. Admittedly, the boundaries are fluid since most tests are conducted on the basis of models, but extreme examples that could be subsumed under the heading of large technical systems highlight what is meant here. For instance, by dealing with individual components nuclear reactors, oil-drilling rigs, chemical productions plants, or even the escape of genetically engineered bacteria can be successfully modeled for the purpose of studying and assessing both their safety and potential hazards. But their systemic nature as well as the environment in which these projects operate are too complex for such modeling procedures. In this sense, technical systems turn into models of themselves: the observation of their functioning, and especially their malfunctioning, on a real scale is required as a basis for further technological development and also for increasing their safety.

Crossing the laboratory's boundaries, stepping out of its protected space that shields society from the risks of scientific experimentation and technical tinkering and the researchers from society's fearful reactions, signals a major institutional change in science. Its established demarcations are becoming blurred, and science itself is turning into a "technique" of applying and implementing knowledge. If the production of knowledge and the implementation of technologies merge to become one process, scientists will inevitably create hazards for which they have not previously obtained the public's consent. There is no institutional framework for this because in the past such joint consultation was simply not required on a politically relevant scale.

The deeper "rationale" underlying the lack of acceptance of new technologies becomes plausible against this backdrop: as we all know, even accepted risks, such as smoking, are judged quite differently from a danger imposed on us by others. Precisely because the "others" can be identified as a group, i.e., highly specialized scientists and engineers, their right of endangering us is being challenged. It is often claimed, however inaccurately, that the public is growing less inclined to accept risks. Rather, it is becoming increasingly opposed to accepting the hazards that the "experts" are creating especially when they claim the right to determine what is the "true" rationality and what are the "appropriate" responses, and when politicians use the same experts' arguments to legitimate policy choices.

The Social-Experimental Character of Large Technical Systems and the Problem of Legitimation

The extension of science beyond the laboratory is by no means ubiquitous; rather, it is limited to a particular class or type. This can be described as the large technical system which is science-based, where the boundaries between research and technological development are fluid. In the public debate nuclear energy has been the focus of attention for a long time and can therefore serve as a paradigmatic case. Time will tell whether the aspects of this technology that have contributed to creating risk awareness will be applicable to other technologies as well.

Possibly the most precarious aspect of this technology is its quasi-experimental character (Krohn and Weingart 1986).[6] Its most prominent German protagonist, Wolf Häfele, speaks of "hypotheticality" which, he claims, is due to the fact that testing by trial and error cannot be used with this technology (Häfele 1974). In other words, it is implemented without scientists knowing in advance what the exact result is going to be. This situation is underlined, for instance, by the premature shutdown of reactors whose construction is a scientific-technological success, but a disaster in economic terms. Science-based technologies, such as nuclear energy, are for their own inherent reasons implemented on an experimental level because the knowledge they incorporate cannot be tested or taken as empirically verified. Whether nuclear power plants pose a risk under real operational conditions, and whether accidents can be controlled under the given administrative and political conditions, and which kinds of impacts are caused by accidents under different climatic conditions, can be correctly ascertained only after an accident has occurred.

Moreover, the technology in question also has "catastrophe potential," that is to say a malfunction may have consequences that are potentially uncontrollable in temporal and spatial terms and therefore with implications for the political and economic spheres as well (Perrow 1984, chs. 2 and 3). Its complexity and its catastrophic potential does not permit real-life testing, as was demonstrated by the Chernobyl disaster (H. Petroski 1982).[7] The implementation of nuclear technologies thus exemplifies a development that Krohn and Weyer have termed "risk transformation," which gives concrete expression to a previously claimed general shift of risks from the scientific-technological system to society (Krohn and Weyer forthcoming).

Nuclear energy also exemplifies the process of increasing technical remoteness. System operation is highly centralized and requires strict controls of access. Its implementation and exploitation involve large amounts of capital expenditure. These requirements and the high danger potential demands a considerable measure of political and administrative commitment. As a strongly research-oriented high technology, its development and operation lie in the hands of highly specialized experts (La Porte 1981). The system is therefore inaccessible to the public.

It is all these aspects taken together, with nuclear energy being an extreme but instructive example, that may explain why the inherent, potential dangers are perceived as imposed, as distinct from those that may be self-imposed.

This impression is reinforced by the fact that large technical systems are especially dependent on political decisions, in several ways. First, such technologies can operate only within a framework of government-funded R&D programs and are usually embedded in a larger scenario of long-term modernization strategies, often concerning infrastructure. Implementation, operation, and control—especially administrative arrangements covering security, technical safety, and potential accidents—far exceed conventional mechanisms of cost calculation and the settlement of civil liability claims. Large technical systems, such as nuclear power plants, require special government guarantees in terms of safety and finance. As the safety of such technology is a matter of public concern, there is an inevitable involvement not only of the scientific-technological community, but also of the political establishment over the legitimation of what decisions will be taken. More than other political decisions, the political decisions involving technological projects have to be legitimated by knowledge as represented by experts and publicly demonstrated by their consensus.[8]

This is demonstrated by a phenomenon resulting from the joining of legitimating experts' opinions with policy decisions characteristic of large technical systems. The implementation of nuclear energy, in particular, illustrates that political controversy over this type of technology had a parallel in scientific controversy. Since the 1970s, the political orchestration of scientific experts has become a fact of political life just as much as the realization that no agreement on the potential risks of new technologies seems possible among experts. Concrete experience ultimately confirmed the suspicions of social scientists, namely that, in its assessment of risks, the public does not and perhaps never did rely on the judgment of scientific and technical experts, but on the credibility and legitimation of the institutions that make the decisions on the implementation of new technologies.

All of the three "big" nuclear disasters—at Windscale in Great Britain, at Harrisburg on Three Mile Island, and at Chernobyl in the Soviet Union—demonstrated that technical experts and political agencies were in a hopeless disarray over how to deal with the aftereffects, and consequently incapable of coping with the political fallout. They seemed bewildered and they lied. Credibility and legitimation were lost, with the result that the public developed its own strategies for obtaining information and fighting the dangers. When, in the aftermath of the Chernobyl disaster, the German Federal Government felt obliged to centralize and standardize radiation measurements, it virtually admitted that it was unable operationally or politically to deal with the sheer volume of measurement and interpretation that was demanded in the weeks after the accident. It is apparent they want to preclude such confusion should there be another accident. It is unlikely, however, that these measures will succeed in restoring the credibility of radiation experts. Information issued by official sources on the hazards of a certain technology is manifestly unconvincing when it is a well-known fact that official policies support this technology, and when, at the same time, the experts themselves are disputing the case.[9]

The events just mentioned are only some of the more spectacular examples of the close collaboration between science, technology, and politics in the implementation of megatechnology. The strict separation of these systems, on which their legitimation once rested, both for purposes of production and as a safeguard against risks, could be reinstated. In the scientific sphere there existed an institutional framework of contained laboratories, an academic dialogue, and a mostly self-referential university system (the proverbial "ivory tower"). These conditions are no longer: the laboratory has become too narrow, and politics is now orchestrating science for its own ends—symbolically, to legitimate policy choices by referring to scientific knowledge, and practically, to achieve progress as defined in technological and economic programs. To the extent that politics has linked itself to the fate of science and technology and vice versa, the institutional arrangement of the production of risks and the provision of safeguards is being jeopardized.[10]

Risk as a Resource and a Result of Social Negotiation Processes

The close interaction that has developed between a science turned technology and politics explains why risks have ceased to be merely a category of scientific calculation and have become a subject for political negotiation. This has recently been addressed in the course of the social-scientific demolition of scientistic risk assessment. Rayner's application of the "grid/group" analysis to the institutionally assignable (and institutionally oriented) perceptions of risk has resulted in a typology of ideal types of attitudinal patterns in relation to risks (Rayner 1984). He stops short of suggesting that these attitudes also underlie behavior patterns that are activated when the issue is about the causation and "imposition" of risks. For Douglas and Wildavsky (1982) this "active" element is still implied because they underscore the functionality of "low probability–high consequence" risks for the mobilization of moral "commitment" for marginal groups, or "sects," as they term them, whose specific problem consists in organizing a permanent membership on a voluntary basis. (Rayner speaks of "egalitarian groups," referring, of course, to social protest movements, public interests groups, and so on.)

The focus is primarily on the perception of risks and the origins of dissent, and less on the "active" aspects of the causation and "imposition" of risks. A closer look at the typology suggests such an interpretation, if only for the reason that it distinguishes central institutions of society from marginal institutions, and can be consistently interpreted in terms of the differentiation assumed here between institutions that cause risks and those that are affected by them. In this sense the grid/group concept can serve both as a sociological account of diverse, incompatible perceptions of risk and as a model explaining the patterns of political negotiations over the cause and definition of risks.

Since the dividing line separates the institutions that produce and implement technology from the institutions that assess technology, the polemics over the risks of new technologies is increasingly turning into a conflict over the design of future technology in which the contenders are trying to influence the genesis of technology. This situation is not new in itself. Every new technology is the result of an innovative process structured by an interplay of technological, economic, political, and cultural motives, interests, and marginal conditions. When a technology finally reaches the marketplace, its prototype has already been changed by a host of regulatory filters which reflect society's experiences with previous technologies, including compensational strategies and safeguards against future risks. Changes in the norms or normative regulations that structure technology are thus indicative—allowing for a certain delay—of changes in the experience with technology (Ropohl et al. 1984).

A study of the history of these regulatory efforts also reveals that past experiences with technology and its risks were variously related either to the technology-producing institutions or to the social institutions which had to integrate the technological developments.[11] It appears that the type of regulation used in a specific case depends on the particular constellation of interests or the political power structure of the day, i.e., on the degree of the institutionalization of the interests concerned. A new technology is perceived as "overpowering" especially when affected interests are not institutionalized because, prior to the advent of the technology in question, they could hardly be affected by it. When the regulation of nuclear technology is considered insufficient in terms of conventional safety standards, or when the discovery of the causes for the destruction of the ozone layer makes the production of fluorinated and chlorinated hydrocarbons suddenly appear unacceptable, the main reason is that new technologies, or the discovery of new hazards of traditional technologies, lead to the development of new constellations of interests. The conflicts over the introduction or continuation of these technologies are, inter alia, an expression of the institutionalization of the newly formed interests as well as an aspect of the bargaining over how to distribute the consequences on the social level. Since the further development of any technology is the result of negotiation processes between groups, the conflicts concerning large technical systems can be interpreted as a result of processes in which the implementation of technology and the identification of the risks involved take place well before the institutionalization of the interests affected by these risks and the allocation of controls or compensational measures. This is documented by the "socio-experimental" character of some technologies.

The general phenomenon is familiar, of course. What is new in the current situation is the differentiation and institutionalization of risk assessment and its increasing authority to define what constitutes a risk and who should be responsible for dealing with the consequences. Risk assessment accelerates the consolidation of political interests via the definition of risks and decisively influences the conflict over their allocation. This provides risk assessment *as a science* with a quasi-representation in the political arena; it is no longer restricted to the (very limited) public impact of scientific discourse and to the equally remote administrative sphere.[12] The identification and definition of risks and the resultant amplification have far-reaching consequences on the behavior of the population. A public communication on the harmful properties of a certain brand of noodles or on carcinogens in beer sets off an avalanche of lawsuits with costly implications for the producers, while a product marked as environmentally safe meets with a positive echo. These examples show that risk determination has become an important political resource: the definition of risks, the singling out of those on whom they may or may not be imposed, the assignment of competencies as to who should be authorized to measure risks, and finally, the allocation of responsibility for their control—all these constitute consequential decisions in a process that resembles a struggle over the distribution of power.

In view of the eminently sensitive, political function of risk or technology assessment, the question arises as to whether they will be capable, by way of scientific rationality and promoting consensus, of delivering a greater measure of legitimation.

The Legitimation Trap of Technology Assessment and Risk Management

Any attempts to avert dangers or eliminate the defects of scientific and technological activities are possible only via an increased complexity of such activities. This is exactly where technology assessment and risk management come in. According to developmental logic, the next step would be to counteract the one-dimensionality of science and technology by increasing the number of parameters, by considering synergetic effects, and by accounting for neglected frames of reference of social systems. The progress achieved in this way is impressive. One anticipated and hoped-for effect, namely that this would also increase the acceptance of the various technologies, has not materialized, however. The reasons are of a logical and epistemic nature, with consequences on the sociological level.

The scientific reasons have been identified for some time, though perhaps not fully acknowledged as limitations. In trying to make complex systems function more safely, control systems are required with an increased level of complexity. Attempts to extend control to environmental systems entail exactly the same infinite regress. Depending on the system in question, there will invariably be a point reached beyond which its safety cannot be further improved. A case in point is the problem of computer viruses. Apparently, the search for a virus in the highly complex programs of NASA is being called off because computer scientists are perplexed by the opaqueness of these programs caused by the vast number of programmers involved and the magnitude of software structures.[13]

There is another, epistemic reason. With technology assessment and risk research now firmly established, risks are being identified even for those technologies whose implementation used to be regarded as unproblematic. Krohn and Weyer have called this phenomenon "risk constitution"—a process that has initiated an ecological communication. In Luhmann's constructivist terms, it is this and only this communication that poses an ecological danger to society, in the shape of "Aufregungsschäden" ("damage through anxieties"; Luhmann 1986a, 1986: 229). Since both the material as well as the temporal frames of reference for technological activities are unlimited and in practice serve the identification of systemic connections, there can be no systematic but only pragmatic limitations for questions about potential impacts. Limitations of the analysis and detection of "risks" (who or what poses a risk to whom or what?) are as wide as the number of observers is great and their viewpoint is large.[14] For this reason alone it is clear that technology assessment and risk research will develop a dynamic force of its own. Risk research lacks built-in checks or curbs and the profession which has made risk definition its business has already been identified with the Inquisition and witch-hunts (Clark 1980).

The discovery and definition of the so-called *Altlasten* (industrial wastes; literally, "old wastes") highlights the situation. As everywhere else in science, the research concerned with this has assumed a dynamic quality of its own and keeps discovering more and more waste. Consequently, more and more aspects of technological activities are being challenged with respect to their legitimation. Apart from the question of the benefits of a new technology, the consideration of costs—i.e., risks and damages it may cause—is becoming increasingly important. And old technologies are no longer immune from technology assessment.

To summarize, many new technologies are being rejected because technology assessment has exposed the potential for an ever-widening spectrum of negative repercussions, and the old technologies are equally becoming alienated from their previous acceptance as more and more of their risks are being uncovered. Science and technology—and politics as well, which is ultimately responsible for the former's implementation or prevention—have all become caught in a legitimation trap. Both the protagonists and their critics have fallen between two schools, those of the old and the new technologies.

To rephrase the dilemma one could say that science provides the only way of making good the damages and eliminating the errors that are of its own making, but in doing so it produces a growing measure of self-addressed distrust. Considered in this way technology assessment and risk assessment are hardly suited to solve the fundamental problems of modern scientific and technological developments and their legitimation. A possible way out of this legitimation trap can be expected only from new political constellations in which legitimation could be obtained for riskful decisions. Alas, no such constellations can be either predicted or expected to emerge at the present time.

NOTES

1. After they, at first, managed to turn neutral findings on human probability judgments into normative evaluations of incompetence (L. Lopes, "The Rhetoric of Irrationality"; paper presented at the Colloquium in Mass Communication, 19 November 1987, Madison, Wisconsin).

2. Taking this further, Beck speaks of a "competition of rationalities" ("*Konkurrenz der Rationalitäten*"); cf. Beck 1986: 76ff.

3. Originally N. Luhmann n.d.

4. On the particular problems of molecular biology and the contextual reduction in genetic engineering see Bonß, Hohlfeld, and Kollek 1988: vol 1., pp. 13f., and vol. 3 pp. 1–6.

5. Significantly, this metaphor has been used by Susan Jaffe (1980: 9–10) in connection with nuclear energy. I am indebted to Carsten Krück and Georg Krücken (1988) for this reference in their seminar paper. For a theoretical treatment of the thesis see Krohn and Weyer 1989.

6. For a systematic analysis of the experimental character of technologies cf. W. Krohn and J. Weyer 1989.

7. As is generally known, the Chernobyl accident was caused by a testing of the safety system.

8. Häfele refers to the natural laws that, for example, underlie the design and operation of a control lever and are sufficiently understood, as opposed to the primary and peripheral conditions required for the application of natural laws, which are not sufficiently understood (Häfele 1974: 312).

9. When this happens, it forcefully brings home the precariousness of the legitimation of quasi-experimental technologies where the government becomes an accomplice in the shifting of risks. It is now known that the dangers from radiation to which U.S. soldiers and the local population were exposed during the testing of the H-bomb on the Bikini atoll were either kept secret or downplayed by the military authorities (*Der Spiegel* 51/1982: 116; quoted in Krohn and Weyer forthcoming); cf. the documentary film "Radio Bikini."

10. On the institutional arrangement of risk calculations and the provision of safeguards, see Beck 1988: 643, where he cites François Ewald's *L'état providence.*

11. For different forms and examples of the assignment of risks see W. van den Daele 1989.

12. This is paralleled by a special form of professionalization; see Dietz and Rycroft 1987;

Jasanoff 1987.

13. Mentioned by Brunnstein in a paper presented at the University of Bielefeld, December 1988.

14. See the recently published report ("One Big Earth System," *Newsweek* 7 November 1988) on James Lovelook's *The Ages of Gaia* as evidence that the scientific discussion on the systemic linkage of all forms of life on this globe has now reached the public, although the topic of "spaceship earth" was introduced into the discussion years ago.

REFERENCES

Beck, U. 1986. *Risikogesellschaft, Auf dem Weg in eine andere Moderne*. Frankfurt.

Beck, U. 1988. "Die Selbstwiderlegung der Bürokratie." *Merkur* 42: 629–46.

Bonß, W.; Hohlfeld, R.; and Kollek, R. 1988. "Risiko und Kontext. Zum Umgang mit den Risiken der Gentechnologie." Foreword to *Tagung der Sektion Wissenschaftsforschung der DGS* (25 November). Dortmund.

Clark, W. 1980. "Witches, Floods, and Wonder Drugs: Historical Perspectives on Risk Management." In *Societal Risk Assessment—How Safe is Safe Enough?*, ed. R. C. Schwing and W. A. Albers, Jr., pp. 287–313. New York, London.

van den Daele, W. 1989. "Kulturelle Bedingungen der Technikkontrolle durch regulative Politik." In *Technik als sozialer Prozeß*, ed. P. Weingart. Frankfurt.

Dietz, T. and Rycroft, R. W. 1987. *The Risk Professionals*. New York.

Douglas, M., and Wildavsky, A. 1982. *Risk and Culture*. Berkeley.

Häfele, W. 1974. "Hypotheticality and the New Challenges: The Pathfinder Role of Nuclear Energy." *Minerva* 12: 303–22.

Jaffe, S. 1980. "All the World is a Laboratory for Nuclear Technology." *These Times* (November), pp. 19–25.

Jasanoff, S. S. 1987. "Contested Boundaries in Policy-Relevant Science." *Social Studies of Science* 17: 195–230.

Krohn, W., and Weingart, P. "Tschernobyl—das größte anzunehmende Experiment." *Kursbuch* 85: 1–25.

Krohn, W., and Weyer, J. 1989. "Die Gesellschaft als Labor, Risikotransformation und Risikokonstitution durch moderne Forschung." In *Riskante Entscheidungen und Katastrophenpotentiale—Elemente einer soziologischen Risikoforschung*, ed. J. Halfmann and K. P. Japp. Opladen.

———. Forthcoming. *Society as Laboratory: The Production of Social Risks by Research.*

Krück, C., and Krücken, G. 1988. "Sicherheit und Kontrolle großer technischer Systeme." Bielefeld; manuscript.

Luhmann, N. 1986. *Ökologische Kommunikation*. Opladen.

————. 1986. *Die Welt als Wille ohne Vorstellung, Die politische Meinung.*

————. 1988. "Wissenschaft." Bielefeld; manuscript.

————. N.d. "Die Moral des Risikos und das Risiko der Moral." Bielefeld; manuscript.

Perrow, C. 1984. *Normal Accidents: Living with High-Risk Technologies.* New York: Basic Books.

Petroski, H. 1982. "When Cracks Become Breakthroughs." *Technology Review* (August/ September), pp. 8–28.

La Porte, T. R. 1981. "On the Design and Management of Nearly Error-Free Organizational Control Systems." In *Accident at Three Mile Island: The Human Dimensions,* ed. D. L. Sills, C. P. Wolf, and V. B. Shelanski, pp. 185–200. Boulder, Colo.: Westview Press.

Rayner, S. 1984. "Disagreeing about Risk: The Institutional Cultures of Risk Managment and Planning for Future Generations." In *Risk Analysis, Institutions and Public Policy,* ed. S. H. Hadden, pp. 150–69. Port Washington.

Ropohl, G.; Schuchardt, W.; and Lauruschkat, H. 1984. *Technische Regeln und Lebensqualität, Analyse technischer Normen und Richtlinien.* Düsseldorf.

Slovic, P. 1987. "Perception of Risks." *Science* 236: 280–85.

Wynne, B. 1982. "Institutional Mythologies and Dual Societies in the Management of Risk." In *The Risk Analysis Controversy. An Institutional Perspective*, ed. H. C. Kunreuther and E. V. Ley, pp. 127–43. Berlin, Heidelberg, New York.

THE GOVERNANCE OF LARGE TECHNICAL SYSTEMS: THE CASE OF TELECOMMUNICATIONS

Volker Schneider
Max Planck Institute for Social Research
Cologne

Societal Governance and the Rediscovery of Institutional Analysis

The varieties of organizational forms societies have developed in their evolution have always been cornerstones of social analysis. From such a perspective institutions ranging from kinship and family structures to state organizations were often conceived as the basic elements of social organization, fulfilling important tasks in societal reproduction and social integration. As a consequence of the behaviorist revolution in the social sciences and the rise of functionalist systems theory, empirical institutions as analytical entities paradoxically lost their importance. In the search for the fundamental forces which are shaping human action, system-functionalism treated institutions as only surface phenomena. When this viewpoint declined in the last decade and when the "sociological deficit" of the new dominant rational choice approach became apparent, the social sciences were in the same way rediscovering the importance of empirical institutional analysis.[1] From this new perspective, institutions are seen as autonomous entities which exist in their own right and play an important role in societal self-regulation. Similar theoretical developments took place in economics, where empirical institutions have been disregarded for a long time. In neoclassic theory, for instance, the fiction of the market as "the natural state" of economic organization was for a long time the dominant perspective.[2] Since the 1970s, however, there is a trend toward the formulation of theories which take into account the variety of organizational forms that historical societies have invented. In this undertaking the concept of governance structures plays a significant role.[3] (Governance is understood as the institutionally structured process of self-regulation of social activities.) A governance perspective has close affinities to cybernetic theory and concepts such as control, regulation, and guidance. A significant difference, however, is that it stresses very much the organizational factors and the institutional embeddedness of social regulation and control. In contrast to system-function analysis, it is less interested in finding general principles of social organization than in understanding the variety, heterogeneity, and complex combinations of institutional control devices. In contrast to early institutionalism, governance theories are firmly based on methodological individualism. Social integration is thus not taken for granted and

I am grateful for the valuable suggestions and criticisms of an earlier version of the paper made by Jürgen Häusler, Renate Mayntz, Fritz W. Scharpf, Frank Thomas, and Raymund Werle. Thanks are also due for the research assistance of Gerda Ehrlenbruch and for the English language guidance of Cynthia Lehmann.

T. R. La Porte (ed.), Social Responses to Large Technical Systems, 19–41.

presupposed. Precisely the factors and conditions which foster or inhibit the evolution of institutional arrangements with different capacities (performance, efficiency, etc.) in societal problem solving constitute a *core problematique* of these approaches.[4]

The Governance of Large Technical Systems as a Regulation of Technical, Economic, and Social Risks

Theories of societal control and self-regulation cannot presuppose the emergence of efficient societal problem solvers and control devices but have to show when and how such devices emerge. From such a perspective the functioning of societal self-regulation cannot be derived from the mere existence of norms and/or interaction rules. Individual and collective actors may, under certain conditions, consciously deviate from norms if they perceive it to be advantageous to do so. Societal self-regulation is thus a very dynamic process where collectivities constantly invent new institutional devices for minimizing the risks of social disintegration. In this respect the recent economic institutionalism focuses on the inherent risks in economic transactions. A central problem is that contractual performance cannot be taken for granted. Neo-institutionalists see actors as basically self-interested and opportunistic. Economic transaction systems thus necessitate governance institutions by which these risks are handled or reduced. Specific non-market forms of economic organization, such as vertically integrated corporate hierarchies or long-term contractual networks between firms, are conceived as institutional responses to transaction problems.[5] Transactions can sometimes be maintained and executed only by hierarchical control, bureaucratic monitoring, and the coordination by a "visible hand." More informal or "mixed" institutional regimes, such as relational exchange networks, obligational contracts, associational structures and informal alliances,[6] may perform similar functions with varying governance costs and regulative effectiveness.

Typical governance problems emerge in Large Technical Systems (LTSs), which may be described as socioeconomic or infrastructural sectors built around a core technology.[7] Establishment and operation of LTSs requires a number of regulative tasks and control processes. The institutional or organizational dimension of such systems can be seen as "governance structures" which guarantee the system's functioning not only with regard to social, economic, and political imperatives but also with regard to technical ones. Functional risks, systems reliability, and the possible negative externalities of LTSs are often not only associated with the way economic and social processes are organized inside the system but with the design of its technological structure as well. This is the way the constitutive artifacts are shaped and interrelated.[8] Some LTSs require special organizational arrangements dealing with particular technical problems or risks associated with the nature of the technologies. For instance, institutional devices have to ensure that the technical system components are fully compatible, that quality standards are met, and that required rules are consistently applied. Security problems and the possibility of negative externalities are further governance problems necessitating institutional regulation.

Like other social configurations, LTSs display a variety of different organizational forms. Recent studies, for example, have pointed to the important role of the state in resource mobilization, planning, management, and operation of LTSs.[9] A number of studies demonstrated that a range of alternative organizational arrangements are feasible. It was shown that private corporations can also perform the role of system builders and/or operators. In addition, resource flows inside LTSs may take the form of market transactions. Many LTSs have rather heterogenous governance structures and incorporate mixtures of market-based transactions, hierarchical technical coordination, and state regulation. Comparative historical analyses of technical systems provide broad empirical evidence of great organizational variety. Probably the most centralized and bureaucratized organization was portrayed in Wittfogel's *Die orientalische Despotie* [*Oriental Despotism*], a study of societies primarily dependent on large-scale irrigation systems. The installation and operation of these systems presupposed the incorporation of the whole society into a centralized administrative hierarchy where governance structures of the LTS and state institutions became indistinguishable. Such centralization would be unthinkable in highly differentiated societies. Although in modern societies the governance structures of certain LTSs intersect with political bureaucracies, there are no cases where they coincide. In advanced industrial societies, LTSs tend to be specialized and differentiated organizational mixtures of markets, contractual networks, hierarchies, and, sometimes, associative structures. The understanding of these organizational forms and their specific relationship to the pure technical dimension of LTSs is an important topic which has not yet been addressed systematically.[10] It is the goal of this paper to examine these relationships in the telecommunications domain to get a deeper understanding of the organizational dimension of large-scale technology.

Emergence and Transformation of Governance Structures in Telecommunications

The following section outlines the organizational development of telecommunications systems in a comparative and historical perspective. The telecommunications development is traced from the invention of the telegraph to the emergence of the telephone in the 1870s and up to the explosive differentiation of the system in the 1970s and 1980s. The article focuses on the emergence and the changes of organizational forms by which telecommunications was governed. It is hoped that such a focused comparison can lead to a sounder understanding of the institutional dimension and its transformations in a given technological system. While some arrangements (such as market competition) transformed rather rapidly into monopolies, other institutional capacities (such as special regulatory policy arrangements) were invented only after decades of telecommunications development.

The technical, economical, and at times political activities related to telegraphy, telephony, and other forms of telecommunications can be understood as the "telecommunication system." Its core function is to enable communication over a geographical distance by means of electric signals. This definition excludes the postal system from telecommunications—although postal communication fulfills similar functions and competes with telecommunications in some areas. We will distinguish the social media of transmission (text, language, and fixed and moving pictures) from the technical means of transmission (wire, microwaves, or laser). The first electric telecommunication systems were the telegraph (invented in the 1840s) and the telephone (invented in the 1870s). Both were based on wire transmission. The invention of wave transmission made the development of radio and television in the 1920s and 1930s possible. The first telewriter was introduced in Germany in 1933. Facsimile transmission also originated in the 1930s. Until the 1960s all of these techniques were based on analog transmission (hertzian waves). With the advent of digital technology in the 1960s and 1970s, a broad range of new telecommunication forms and techniques emerged. The most important ones are the different forms of data transmission, such as specialized data networks, videotex, and teletex. Facsimile became an affordable means of business communication only through digitalization.

Parallel to the emergence and the succession of technological innovations, the telecommunication system diversified into an array of coexisting technologies and services (see table 1). During the same period, the network size of telephony grew rapidly. Whereas telephony was limited to local areas until the turn of the century, in the following decades it was quickly integrated at the national level. Since World War II telephony and the other telecom networks expanded to an unprecedented degree and were integrated into a veritable "world system."

TABLE 1

The Diversification of Telecommunication Services

Period	Services
1847–77	Telegraph, Telephone
1930	Telegraph, Telephone, Telex, Facsimile
1970	Telegraph, Telephone, Telex, Facsimile, Data, Mobile telephone
1980	Telegraph, Telephone, Telex, Facsimile, Data, Mobile telephone, Videotex, Teletex, Electronic mail, Telemetry, Packet switching, Video conferences, Video telephony

The usage patterns of telecom service also developed in stages. In beginning of telegraphy the usage was often restricted to the military and telephony was in the early decades almost exclusively restricted to professional communities. It is only during the last twenty or thirty years that telephony has penetrated the residential arena and become an almost universal service.

Varying Governance Structures in Telecommunications

An important question of telecommunications development is how the institutional forms by which these systems are governed are related to the system's growth, expansion, diffusion, and diversification. These institutions distribute tasks and rights in development, establishment, and operation of the system. Such institutional structures determine responsibility, for instance, for R&D activities, for the operation of the system, and for the coordination of production. They also organize the way negative externalities are prevented or reduced.

Governance structures in telecommunications differ in three primary ways: they differ in the degree to which the system is controlled by the political sphere (*public control*); they vary in the degree to which the flow of production from raw material to the consumer is separated or integrated (*vertical integration*); and they vary in the extent to which all direct and indirect competitors are incorporated into one organizational unit (*horizontal integration*).

Existing institutional arrangements can be considered as a combination of these three institutional facets. An overview of the organizational variety in telegraphy and telephony across ten countries is depicted in figure 1.[11] At least five typical governance regimes can be distinguished:

(1) a public monopoly, where network/service provision and systems operation is provided by a governmental department or by a public corporation;
(2) a monopoly held by private firms which receive concessions from the state;
(3) a monopoly held by a private firm which is combined with governmental public utility regulation;
(4) an unregulated private monopoly;
(5) competition between different firms and/or public institutions.

The historical comparison of institutional development paths provides some important results. National systems started from several organizational starting points and followed different development paths in their subsequent growth. The variations, however, differed between telegraphy and telephony. In telegraphy most countries started from the very beginning with a state monopoly. In telephony only two countries started from a public monopoly (West Germany and Japan). Six countries started with institutional arrangements that regulated a private, licensed monopoly (Austria, France, Italy, Netherlands, Switzerland, and the United Kingdom). Only two started with or moved through a period of free competition (Sweden and the United States). After some time nearly all systems

FIGURE 1

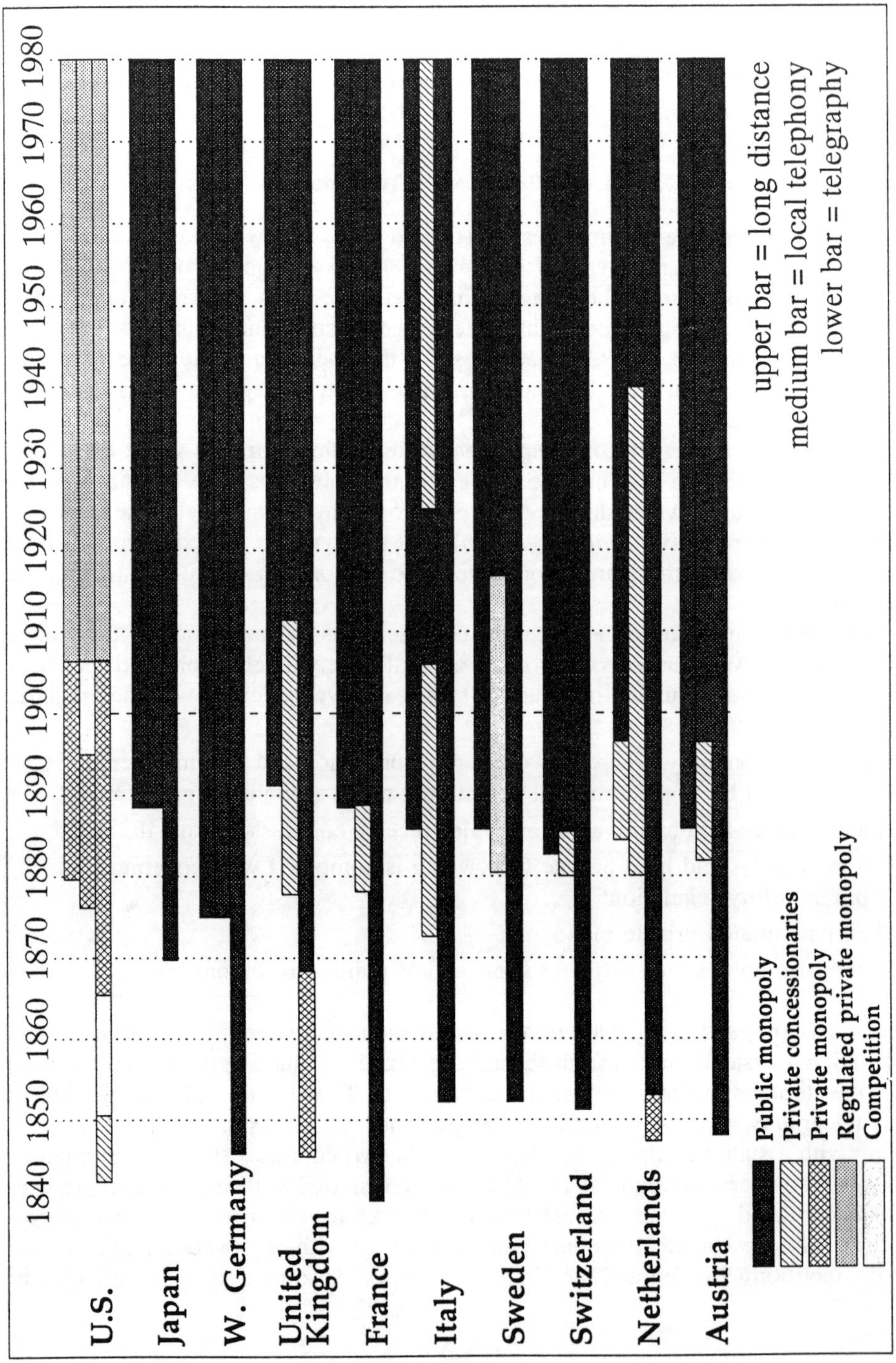
1840 1850 1860 1870 1880 1890 1900 1910 1920 1930 1940 1950 1960 1970 1980
U.S.
Japan
W. Germany
United Kingdom
France
Italy
Sweden
Switzerland
Netherlands
Austria
Public monopoly
Private concessionaries
Private monopoly
Regulated private monopoly
Competition
upper bar = long distance
medium bar = local telephony
lower bar = telegraphy

converged in a single "organizational paradigm" which lasted from the beginning of the 20th century to the late 1970s or early 1980s.

The organizational paradigm in telecommunications that existed in the majority of countries (with some minor variations)[12] can be described by a few organizational variables. It combined strong public control in the form of a public administration (or public corporation), a quasi-vertical integration where the operation and terminal provision was controlled by a hierarchy, and in most countries a completely horizontal integration of post, telegraphy, and telephony. In this model public ownership fulfilled and integrated several functions: planning, capital mobilization, operation, standardization, R&D etc., and regulation. Only the U.S. model is a special case: initial competition converged quickly toward a private monopoly which, however, became increasingly regulated by the state. Furthermore, the U.S. model entailed complete vertical integration from production, operation, to the terminal provision.

A look at the variation of the different governance structures during the "old order" (from early in the 20th century to the 1970s—before the radical change emerged!) shows that only a few of all the theoretically possible combinations of the three institutional facets have been used.[13]

The *public control* aspect can be differentiated into four variants. The strongest and most directly politically controlled organizational form was a telecom administration operated as an ordinary governmental department based on the general state budget. More relaxed is political control in public corporations with a proper, or at least a relatively autonomous, budget. Private corporations subjected to public utility or antitrust laws or other forms of governmental regulation can be considered as even less controlled by the political sphere. The lowest degree of political control exists in a monopoly held by a private corporation or in a situation of free competition between private firms which are only subject to corporate and private law.

The organizational *horizontal integration* facet can also be divided into four structural elements. The most complete form of integration is the typical PTT monopoly, where

TABLE 2

Organizational Facets of Governance in Telecommunications

Political Control	Horizontal Integration	Vertical Integration
Governmental department	PTT	Complete integration
Public corporation	Telecom and post separated	Manufacturer separated
Regulated private firm	Competition between services	R&D, manufacturer, and terminal provision separated
Unregulated private firm	Free competition; complete horizontal integration	Liberalization of all levels

post, telegraphy, and the telephone are integrated under one roof. A lower degree of integration is the separation and competition between the different subsystems of telecommunications (telegraphy, telephony, other services). Horizontal integration is even lower when competition within subsystems are allowed—for example, inside of telephony. The horizontally most differentiated system is perhaps the current Japanese model, where competition is allowed on almost every level (except within the postal system, as shown in table 3).

The *vertical integration* facet can be traced from a system in which the functions of R&D, manufacturing, operation, and equipment provision are integrated in one organizational unit (complete backward and forward integration). The other extreme is an open and free market, in which all these functions are provided by different organizations exchanging goods and services via market interfaces. A typical intermediate form of vertical integration is the so-called "quasi-integration model," where planning, systems operation, and terminal provision are integrated in one organizational unit. Manufacturing, however, is left to a small set of private firms which are tightly linked to the administration by long-term contracts. A well-known example of such an arrangement is provided by the defense industry. A more differentiated model would also leave the provision of terminal equipment to private industry. Within the most differentiated system, every phase in the vertical production stream is provided by specialized actors and even service provision is differentiated between "basic services" and "value-added services." The role of the traditional network operator is there reduced to that of a mere "common carrier."

The combination of different institutional facets allows the disposition of traditional national governance structures of telecommunications in a three-dimensional space. Figure 2 shows the "traditional organizational paradigm" in the lower left corner. At the end of the 1970s and the 1980s some of the telecom systems began to change. The direction of institutional change is indicated by the arrows pointing to the upper right corner. This new model combines thus a general decrease in political control and in horizontal and vertical integration.

Explaining Institutional Convergence and Change

The cross-national comparison of long-term institutional development processes shows that several organizational transformations had taken place which can be specified by a small set of variables and dimensions. But how can these institutional changes be explained? In order to understand the emergence of the traditional regulatory model, it is helpful to distinguish between four institutional transformations: first, the transformation from initial *competition* (in some places) to a *monopoly;* second, the emergence of a high degree of *public regulation* either by *public ownership* or by *regulatory commissions;* third, the trend toward more *financial autonomy* in organizational forms that were subjected to direct political control; and fourth, the dominance of *quasi-vertical integration* through long-term contracts with "family" or "court" suppliers.

There are several reasons for the first transformation—*monopolization*—and for the

FIGURE 2

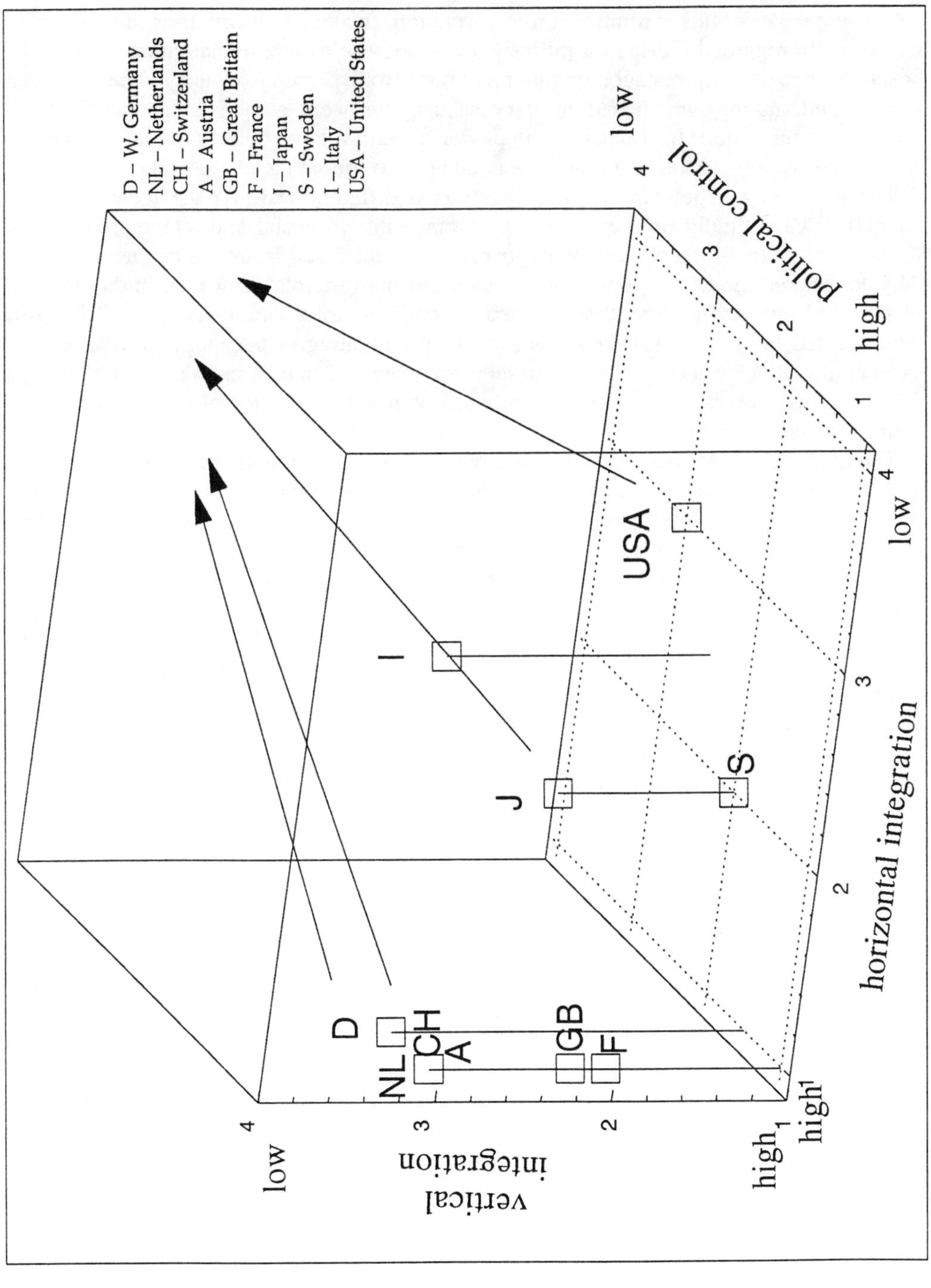
D – W. Germany
NL – Netherlands
CH – Switzerland
A – Austria
GB – Great Britain
F – France
J – Japan
S – Sweden
I – Italy
USA – United States
political control
horizontal integration
vertical integration
low
high
1
2
3
4
D
NL
CH
A
GB
F
J
S
I
USA

existence of public monopolies. In most cases, it was simply the interest of governments in maintaining their rights of sovereignty, political power, and financial revenues that was responsible for this institutional transformation. In almost all countries, the telegraph was initially regarded either as a military technology or as a potential competitor to the postal service in long-distance communications. Both perceptions led to the idea that private undertakings threatened military security or were a menace to the financial revenues of the state.[14] In countries with powerful central governments, private initiatives were quickly stopped and telegraphy was taken into the hands of the state.

The monopoly in telephony came about in a different way. At the local level the telephone was initially not perceived as a competitor to postal and telegraph services. In the early years of telephony, many governments abstained from this business because they considered the technology to be irrelevant or unprofitable. However, in the position of postal monopoly holders, they granted concessions to private companies. The firms then secured monopoly positions from the very beginning or became dominant after a short competitive period by buying out their opponents. The inherent nature of telephone networks—i.e., economies of scale combined with network externalities[15]—then transformed initial competition sooner or later into a monopoly.

The evolution in long-distance telephony was entirely different. At this level the new system competed directly with the postal services and telegraphy. When governments as postal monopoly holders realized this challenge to their revenues, they extended their legally protected monopoly positions to long-distance telephone networks. Later, the ownership of trunk lines was often used as a leverage to get local telephony under control.

The emergence of *public control over private monopolies* resulted from this technology's important role in military communication (and thus national security) and from the fact that in almost all countries the institutional structure of telegraphy had been historically preceded by a state monopoly in the postal services. The perception of telegraphy as a threat to the state revenues in the postal domain was thus reason enough to get this technology under public control. It is striking that in most countries telegraphy started with a public monopoly.

The typical example of a private monopoly in telephony is provided by the United States. But this system also came under increasing political control in the form of indirect state regulation. The most convincing explanation for a general tendency toward a political control of telecoms seems to be that private monopolies inevitably tend to have comparatively high telephone rates (monopoly pricing) and to discriminate between user groups.[16] Monopoly behavior created public discontent, and the government or the municipalities were therefore urged to intervene either to regulate the system or take it over into public ownership. As a result three basic types of public governance structures evolved: public ownership (Austria, Germany, France, United Kingdom, Netherlands, Sweden, Switzerland, Japan); franchising with regulatory conditions (Italy); and rate-of-return regulation (United States).[17]

A third transformation—*increasing financial and planning autonomy of the national postal or telecom administrations*—took place within the group of public monopolies.

This striving for autonomy may be explained by the strains between political and technical rationality that emerged in entirely government-controlled telecommunications systems—i.e., governmental control is subjected to the logic of politics, which in parliamentarian institutional arrangements is exposed to rather short-term pressures. Technical innovation, planning, and investment in telecommunications, in contrast, are governed more by a "technical rationality" that is determined mainly by the speed of technological development. During the period of the old telecom technology, the product cycles in telecommunications were extended over a relatively long period. A rough indicator for such cycles is technical obsolescence, generally indicated by depreciation schedules. In telecommunications these lasted up to thirty-five years during the pre-digital age (in Germany ten to twenty years, in the United States twenty to thirty-five years; OECD 1973). It is obvious that such long-term planning created problems within normal parliamentarian budgetary control practice. Furthermore, completely publicly controlled horizontal integration with no financial autonomy often led governments to exploit telecommunications revenues for their general expenditures. Instead of investing into network expansion, the state would use the money to finance its (often chronic) deficit. In countries such as France and Japan, this led to serious crises in telephone development. In other countries these conflicts between technical and political timing were resolved by the separation of the telecommunications administration from the governmental bureaucracy. One example is the transformation of PTTs into public corporations, or at least into some mixture of a governmental department and a public corporation with an autonomous or semi-autonomous budget.

A fourth transformation is the *quasi-vertical integration* of the system. From the beginning of this century the industrial organization of telecommunications systems converged in an organizational model that left manufacturing to a relatively small set of private firms which were linked by long-term contracts to the telecommunications administration. Systems operation and terminal provision were reserved for the telecom administration. Before the digital revolution began to transform the market structures in telecommunications, this "court supplier" model could be observed in almost all countries. Only the United States and Sweden were exceptions, in which a completely vertically integrated system emerged.

The quasi-vertical integration model had some important advantages over complete integration. The existence of several "suppliers to the court" reduced the telecommunications administration's risk of becoming dependent on one company because competition was possible among members of the "family." The family was, however, always small enough to ensure sufficient economies of scale. The supply relations were relatively stable, because the "asset specificity" of investment was very high. This created high entry barriers for new entrants (sunk costs) and led to stable "family structures." Quasi-vertical integration thus provided the advantages of integrative security, limited competition, and economies of scale.

The Disintegration of the Old Model

Until the end of the 1970s most of the observed telecommunications systems were organized as publicly administered monopolies which were financially rather autonomous (compared to other state administrations). And most of these systems could be characterized as vertically quasi-integrated domains. This organizational configuration began to change significantly in the 1980s. In a number of countries, a new trend emerged: there was a *decrease in horizontal and vertical integration* and a tendency toward *privatization and less governmental control.*

The first radical institutional reforms were undertaken by the United States, the United Kingdom, and Japan. The Netherlands, Sweden, West Germany, and France undertook similar reforms during the following years. The governance structures in telecommunications that existed in 1985 within the countries under investigation are summarized in table 3.

Why did institutional configurations which proved to be fairly stable over a long period all begin to change in a rather short time? In the following section, I identify some causal factors and processes that may explain these new institutional transformations.

TABLE 3

Telecom Governance Structures in 1985

	A	CH	F	D	GB	I	J	NL	S	USA
Relation with postal service	PTT	PTT	PTT	PTT	S	S	S	PTT	S	S
Terminal equipment										
Telephone set	M	M	PL	M	L	L*	L	L	L	L
PABX	L	M	L	L	L	L*	L	L	M	L
Telex	?	M	L	L	L	L*	L	L	L	L
Modem	M	PL	L	PL	L	L*	L	L	PL	L
Basic service network										
Local	M	M	M	M	RC*	M	RC*	M	M	PM
Domestic	M	M	M	M	RC*	M	RC	M	M	FC
International	M	M	M	M	RC*	PM	RC*	M	M	RC*
Leased circuits										
Resale allowed	n	n	n	n	y	n	y	y	n	y
Connection to *public networks*	n	n	n	n	y	?	y	y	n	y

S = Post and telecom separated
M = Monopoly
PM = Private monopoly
L = Liberalized
PL = Partially liberalized
RC = Regulated competition
* = Restricted
FC = Free competition
y = yes
n = no

Source: OECD 1987.

There are at least two classes of explanations for this institutional transformation. On the one hand, institutional change could be related to policies of deregulation which were formulated and implemented in a number of countries, especially where conservative political forces were in power.[18] In this scenario institutional change would be the reflection of a significant ideological transformation (emergence of neo-conservatism or revitalization of liberalism). Empirical evidence for this view is provided by some countries in which conservative movements came to power. In these countries deregulatory measures were components of a general political program that planned a cutback in state intervention and was thus not restricted to a specific societal sector. Thatcherism may explain the deregulation in Britain to some degree and the neo-conservative revolution under Reagan may explain parts of the U.S. development (although the first deregulation measures can be traced back to the Carter administration; the first important liberalization measures in U.S. telecommunications were undertaken in the 1970s).

Another explanation adopts a historically more extended view. In this scenario institutional change is not just a policy outcome representing situational constellations of interests but a long-term adaptation responding to important technological and economic transformations that occurred during the last twenty years. The new technical revolution is of particular importance, for the development of digital technology radically changed the technological basis of telecommunications. It sped up innovation, abbreviated the product cycles, and lead to the integration of branches which formerly were rather sharply divided on the basis of different technologies. New electronic technologies became more generally applicable and the formerly rather high "asset specificity" in telecommunications decreased significantly. This meant lower sunk costs for new entrants and lower production switching costs. As the basic technology for different sectors became the same, the blurring sectoral boundaries made it very difficult to preserve the institutional differentiation between the telecom sector (highly regulated) and the computer domain (unregulated).

The break-up of the equipment and service markets in telecommunications, as the most obvious sign of institutional change, has to be seen mainly as an outcome a technological revolution. An equally important development was that new forms of transmission eroded the natural monopoly of traditional telephone networks.[19] This cluster of technical and economic transformations created strong pressures toward a global change of the existing institutional arrangements. The more the competition in equipment markets increased, the more the manufacturers felt their asymmetric dependency on the telecom administrations. Industry became not only interested in a liberalization of terminal markets but also of the markets for "value-added services." Telecommunications network operators tended to be reduced to mere "common carriers."

In many countries restricted institutional changes such as liberalization in some market segments began to show ramifications in other institutional domains. As soon as the markets were opened for private competitors, the industry was not satisfied by its achievements but became increasingly interested in a radical reduction (at least at the financial level) of direct political control in telecommunications. The aim was to prevent cross-subsidization and the misuse of regulatory power—an important condition for fair competition in the telecom markets. The old model thus was transformed quite rapidly into a rather consistent new one. The liberalization of telecom markets, the institutional separation of telecommunications and postal services, and the functional separation between operation, terminal provision, and regulation could thus be seen as a new organizational paradigm corresponding to the new technological and economic situation.

The relationship between situational and structural challenges and institutional responses, however, should not be seen as too tightly coupled. Societal problems and crisis situations do not necessarily lead to institutional innovation and transformation. Existing institutions are sometimes flexible enough to preserve their structure but to adapt to new requirements by the mobilization of internal slacks and hidden resources. The technological pressures in telecommunications could, for instance, have resulted in much stronger resistance from the traditional PTTs defending their traditional positions if their potential for institutional innovation or adaptation had not been seriously exhausted. The growing differentiation in communications needs and the rapid technological change began to overcharge large bureaucratic PTTs. Increasingly criticized for being unable to serve the specialized communications needs, they had to accept the entry of new industrial actors into the telecom domain. Once they had entered the market, the new competitors started to fight for fair competition. This meant an attack on the fortress of the traditional PTT. A radical reorganization of existing telecom institutions was the only response consistent with the new situation.

Indigenous technological challenges in each country were in a way amplified by exogenous pressures in an increasingly interdependent and competitive world economy. Deregulation and institutional changes in a number of countries were therefore not only a result of synchronous adaptation by national institutions to technical and economic changes, but also the outcome of an international "infection process." Although each country experienced domestic pressures for institutional change, the first move toward institutional change in one country put other countries under strong pressure to follow. Seen from a trade perspective this meant that liberalization in one country without similar measures in all other countries would lead to competitive disadvantages. The "first mover" in deregulation is thus strongly motivated to support liberalizers in other countries. A similar logic applies in an industrial policy perspective. From such a viewpoint regulations are seen as competitive burdens for the domestic industry. Interdependencies in trade and industrial policy thus create deregulatory chain reactions. In fact, the first move in U.S deregulation of telecoms had such global ramifications. It amplified the pressure toward institutional change, especially in European countries. AT&T's divestiture not only opened U.S. telecommunications to new competitors from the computer business but also liberated AT&T from some old restrictions and enabled it to move into the computer business. These dynamics created complex policy alliances at the international level. The U.S. trade administration became motivated to engage in negotiations with the highly regulated European countries in order to create fair trade conditions. The liberated U.S. multinationals were interested in supporting the alliance of deregulators in Europe. This mobilized European industrial policymakers such as the European Community, who realized that European industry would lose the industrial battle if the European governments waited too long to readjust their regulatory environments. It became clear that the old order would create serious obstacles for innovation in the new communications technologies.

To sum up, the institutional transformations are not short-term policies revisable by a change in government but rather long-term adaptations to structural changes in technology and the world economy. This means that all advanced industrial countries will follow the general trend—with greater or smaller variations and with greater or lesser speed. Variations around the new trend and differences in the speed of adaptation to the new requirements seem to reflect not only unequal technological development in different countries but also a variation in the willingness and the capacity for collective action to implement institutional change.

The willingness of different countries for institutional change seemed to be a function of two different factors:

1) Traditional organizational configurations differed quite significantly in their performance, i.e., in the degree to which they satisfied their customers' needs (reliability, quality, tariffs, innovation, etc.). This created important differences between countries in the constellated social support for or opposition to institutional reform.[20]

2) The countries differed in their capacity for collective action toward institutional reform. This ability is, on the one hand. strongly dependent on political institutional structures (such as decision-making structures) and decision styles[21] in the different countries, and on the other hand, tightly linked to the power relations between the relevant sociopolitical actors.[22] Rather centralized countries where the power of the legislative and executive branches in policy formulation and implementation is much more concentrated seem to be much better positioned for radical institutional reforms than countries where decision-making and implementation power is distributed among different levels and sectors.

The Emergence of Security Problems, Negative Externalities, and their Regulation

In the preceding sections the description of governance structures was mainly restricted to the internal (economic, social and political) organization of the telecommunications sector. A further question is how the governance structures are related to the control of negative externalities imposed by telecommunications on its social environment. Externality and security problems created by other large technologies such as steam boilers or nuclear power stations were of minor importance for telecommunications. Only a few regulations have been related to these technologies. The first problems emerged during the 1890s when the high-voltage electrical industry came into conflict with low-voltage telegraphy and telephony. The existence of high currents (lighting or street railway circuits) in the vicinity of telephone lines sometimes damaged entire telephone systems. In order to deal with these conflicts an array of technical security regulations had to be developed. In the process of policy formation and the subsequent allocation of regulative burdens, business associations, and semi-public institutions were often used as mechanisms for interest intermediation.[23]

An important aspect of traditional governance was that the regulation of economic activities and the control of technical risks had been incorporated into the same institutional structure. The standardization of network equipment, for instance, as an instrument to ensure network integrity, was provided by the system builder and system operator. In a number of countries a unitary and standardized equipment technology approach was applied in order to increase reliability and economize technical maintenance. However, new technological revolutions rendered long-term planning almost impossible and the liberalization of telecommunications markets forced the telecom administrations to abandon their old approach. The demand for competitive fairness finally pressed the governments to an institutional separation of standardization functions, which are now provided by specialized public, semi-public, or private standardization bodies. At the same time, the internationalization of markets and network interconnections shifted standardization processes more and more into the international arena. An explosive proliferation of standardization bodies in governmental and nongovernmental international organizations is a clear indicator of this institutional transformation process.

Other, more modern security problems emerged in the 1980s. As the telecommunications networks proliferated and expanded, communications became increasingly dependent on the reliability of telecommunications networks and services. In contemporary societies a number of economic sectors depend to such a degree upon telecom networks that they could not survive more than a few days without them. The telecommunications system's role in both the economic stability and vulnerability of modern societies has become more significant—it is, perhaps, becoming a kind of "Achilles' heel" of the information society.[24] Governance structures and control technologies are responsive to these transformations, and new procedural and institutional arrangements are being set up to cope with these security and reliability problems. Many of these new problems are processed within associative or semi-public organizations, where firms, the public administration, and business associations cooperate in the development of security standards.

Another negative externality emerged through the change to digital transmission and the extended use of telecommunications for data communications and transactional services. The technical possibility that large numbers of communication acts and transactions (e.g., telebanking, teleshopping) can be recorded and analyzed with very low costs is now increasingly seen as a threat to privacy. In the meantime a number of countries responded to this challenge by legislating data protection laws affecting not only information technology such as computers but also telecommunications networks. Table 4 shows the countries which already have such legislation or are in the process of introducing it.

The emergence of new externalities in the development of telecommunications creates the need for new instruments of institutional control and leads to some complex developments in governmental intervention. The trend toward less direct governmental control at the level of economic governance seems to be accompanied by a trend toward more externality regulation by the state. This, in turn, leads to new forms and procedures of regulatory state interventions,[27] as well as new mixtures of self-regulatory arrangements between private and public actors.

TABLE 4

Data Protection in Telecommunications

	CH	F	D	GB	J	S	USA
Data protection laws							
In force		1978	1978	1984		1973	1974
Draft law	1983				1988		
Data protection in telecommunications							
Telephone billing	x	x	d		x		
Videotex	d	x	x	x		x	o
Electronic mail	x	x	x				
Telemetry			x				

x = Legislation in force
d = Draft regulation
o = codes of conduct

Source: Gebhardt 1989.

NOTES

1. For overviews on different currents of neo-institutionalism in economics, sociology and political science see Williamson 1985; March and Olson 1984; Shepsle 1989.

2. For a critique of the neglect of non-market coordination regimes of economic activities in economic theory see Arrow 1971.

3. See Williamson 1979, 1986; Hollingsworth and Lindberg 1985; Lindberg, Campbell, and Hollingsworth forthcoming; Crew and Kleindorfer 1985; Schmitter 1989; and Bauer 1988.

4. In the developmental map of social theory building, as Hyden (1988) suggests, governance may possibly become a new paradigm, replacing the existing theories of public and collective choice by providing a conceptual approach of social integration based on a mixture of institutionalism and rational choice. Hyden maps the succession of theoretical paradigms as follows:

	INTEGRATION		
1960s			1990s
	Structural functionalism	Governance	
STRUCTURE			ACTION
	Neo-Marxist political economy	Public and collective choice	
1970s			1980s
	CONFLICT		

5. Such risks are always associated with specific situational constraints and include problems in quality measurement, investment idiosyncracies, and the danger for investors of being locked into high "asset specificities" (for example, investments in dedicated equipment or specially trained staff; Williamson 1986). In a discussion of governance structures in the context of new information technologies, Child (1987) mentions other risks involved in the organization of economic activities, e.g., demand risks, innovation risks, efficiency risks, and asymmetric dependency. Such a constellation often leads to vertical integration in order "to prevent a firm's being held up by another contracting party in the position of being able to cause the firm large losses by altering the terms of the agreement at a strategic moment" (North 1983: 38).

6. This list of governance mechanisms is neither systematic nor exhaustive. Categorizations have been provided along a number of different dimensions. Williamson (1985) identifies five forms ranging from the completely integrated firm and the market, according to the frequency of contracting. Ouchi (1980) identified markets, hierarchies, and clans as different forms of governance. To this threefold typology Boisot and Child (1988) added the "fief" mode. Schmitter and Streeck (1986) and Hollingsworth and Lindberg (1986) differentiate between the forms market, state, community, and association. More recently Lindberg, Campbell, and Hollingsworth employed a typology which includes markets, monitoring networks, obligational networks, promotional networks, hierarchies, and associations. Schmitter (1989) distinguishes between spontaneously equilibrating, self-enforcing, and externally enforced means of governance, on the one hand, and governance types which are either autonomously chosen by rational individuals or collectively bargained by social actors, on the other. Within this two-dimensional space six modes of governance are distinguished: markets, alliances, private hierarchies, communities, networks, and public authorities.

7. See Joerges 1988; Mayntz 1988.

8. The combination of different technologies, with different modules, operating intensity, and interrelatedness, produces a variety of structural differences. An example of a dimension of this technological structure is the distinction of loosely vs. tightly coupled (Perrow 1984).

9. See the studies in Mayntz and Hughes 1988. For the role of the state in the introduction and development of advanced communications technologies (videotex) see Schneider, Vedel, and Miller forthcoming.

10. Only a few references to the social organization of LTSs can be found in the literature. Hughes (1987: 53) states that LTSs would "increasingly . . . incorporate environment into the system, thereby eliminating sources of uncertainty, such as a once free market." Two pages later he concludes that LTSs "mostly prefer hierarchy."

11. The comparison is based on the following literature: Bauer and Latzer 1988; Bertho-Lavenir 1988; Blankart 1984; Bottiglieri 1987; Brock 1982; Cawson et al. 1990; Dang-Nguyen 1985, 1986; Foreman-Peck and Manning 1988; Galambos 1988; Grande 1989; Granstrand and Sigurdson 1985; Harris 1989; Heuermann, Neumann, and Schnöring 1986; Holecombe 1911; Kaijser 1987; Lehmruch et al. 1988; Libois 1983; Neumann 1987; Snow 1982; Thomas 1988; Tomita 1984; Vedel 1988; Werle 1990; and Wieland 1985, 1988.

12. A statistical report of AT&T in 1963 counted among the thirty-five European countries twenty-nine government-controlled national telecommunication systems, five countries with mixed regimes, and two countries with exclusively private telecommunications organizations.

13. An "experiment of thought" based on "facet theory" (Canter 1985) suggests sixty-four theoretically possible organizational facet combinations. In reality, however, only a handful of these facets are used.

14. The postal service was a state monopoly in almost every country. According to Noam (1989: 258) "the postal system was a major source of revenue, just at the time when absolutist European rulers had insatiable needs for money. This goose with its golden eggs was ardently protected through the centuries against encroachment by private competitors and by other states. When the telegraph and later the telephone emerged in the nineteenth century, they were rapidly integrated into the postal monopoly system and guarded by the same protective policies."

15. For the economics of network externalities and their effects on market structures see Katz and Shapiro 1985.

16. There was, for instance, the tendency to expand the network only in areas which were considered to be particularly profitable.

17. For the discussion of rate-of-return regulation as a governance structure see Crew and Kleindorfer 1985.

18. For a discussion and critique see Hills 1989.

19. For this argument see Galambos 1988: 146.

20. For a comparison of tariff differences between the countries see Mitchell 1983.

21. For a study of the effects of different decision rules and styles on the capacity of collective action see Scharpf 1989.

22. For instance Lehmbruch et al. (1988) and Grande (1989) attribute the unequal development in institutional change between Great Britain and West Germany in large part to the higher capacity for collective political action of the more centralized British state structures (i.e., a stronger executive) and the less centralized society (i.e., less centralized and powerfully organized interests). For the general structure of this argument see also Katzenstein 1988.

23. Holecombe (1911) provides an overview on how these security problems were regulated quite differently within the European countries, depending variously on the strength of the different branches of the electrical industry, the position of the telephone administration, and the existence of intermediating institutions.

24. For a case study on this new dependency and vulnerability see Takanashi et al. 1988.

25. An example is the prevention of unauthorized access to telecom networks. This problem can be illustrated by the introduction of the German videotex system. Most of the tele-account services in German videotex offer not only withdrawal of funds but also financial transfers and

standing orders. Abuse of the service is prevented by a special system of personal identification and special transaction codes which was developed by German banks within their peak association (*Zentraler Kreditausschuß*).

26. For a comparison of data protection policies that are related to telecommunications see Gebhardt 1989. For a general comparative study of data protection policies see OECD 1989 and Bennett 1988.

27. An important new form of technological system governance is technology assessment (see Gibbons and Gwin 1985).

REFERENCES

Arrow, K. A. 1971. "The Organization of Economic Activity. Issues Pertinent to the Choice of Market versus Non-market Allocation." In *Public Expenditures and Policy Analysis*, ed. R. H. Haveman and J. Margolis, pp. 59–73. Chicago: Markham Publishing.

Bauer, J. M. 1988. *Alternative Governance Structures for Infrastructure Sectors: Notes on an Institutional Design Problem.* EUI Colloquium Papers. Florence: Badia Fiesolana.

Bauer, J., and Latzer, M. 1988. "Telecommunications in Austria." In Müller and Foreman-Peck 1988, pp. 53–84.

Beniger, J. R. 1986. *The Control Revolution. Technological and Economic Origins of the Information Society.* Cambridge, Mass.: Harvard University Press.

Bennett, C. J. 1988. "Regulating the Computer: Comparing Policy Instruments in Europe and the United States." *European Journal of Political Research* 16: 437–66.

Bertho-Lavenir, C. 1988. "The Telephone in France 1879 to 1979: National Characteristics and International Influences." In Mayntz and Hughes, eds., 1988, pp. 155–77.

Blankart, C. B. 1984. "Das Wachstum des Postmonopols [in der Schweiz]: Ein wirtschaftshistorischer Beitrag zur ökonomischen Theorie der Verwaltung." *Jahrbuch für Neue Politische Ökonomie* 3: 172–97.

Boisot, M., and Child, J. 1988. "The Iron Law of Fiefs: Bureaucratic Failure and the Problem of Governance in the Chinese Economic Reforms." *Administrative Science Quarterly* 33: 507–27.

Bottiglieri, B. 1987. *STET. Strategie e struttura delle telecommunicazioni.* Milano: Franco Angeli.

Bowman, J. R. 1989. "Transaction Costs and Politics." *Archive of European Sociology* 30: 150–68.

Brock, G. W. 1982. *The Telecommunications Industry: The Dynamics of Market Structure.* Cambridge, Mass.: Harvard University Press.

Cawson, A.; Morgan, K.; Holmes, S.; Stevens, A.; and Webber, D. 1990. *Hostile Brothers: Competition and Closure in the European Electronics Industry.* Oxford: Clarendon Press.

Canter, D., ed. 1985. *Facet Theory: Approaches to Social Research.* New York: Springer-Verlag.

Child, J. 1987. "Information Technology, Organization, and the Response to Strategic Challenges." *California Management Review* 30.

Crew, M., and Kleindorfer, P. R. 1985. "Governance Structures for Natural Monopoly: A Comparative Institutional Assessment." *Journal of Behavioural Economics* 14: 117–40.

Dang-Nguyen, G. 1985. "Telecommunications: A Challenge to the Old Order." In *Europe and the New Technologies*, ed. M. Sharp. London: Pinter.

————. 1986. "État et Enterprise Publique: Les PTT Européens et leur Environnement (1)." *Revue des Politiques et Management* 1: 119–52.

Diebold GmbH, ed. 1971. "Bedeutung und Entwicklung der automatischen Datenverarbeitung." *Computer und Angestellte* I (Frankfurt: EVA).

DIW, ed. 1987. "Zur Neuordnung der Fernmeldemärkte." *Wochenbericht* 54, 37: 493.

Du Boff, R. B. 1984. "The Telegraph in Nineteenth-Century America: Technology and Monopoly." *Comparative Studies in Society and History* 26, 4: 571–86.

Foreman-Peck, J., and Manning, D. 1988. "Telecommunications in the United Kingdom." In Müller and Foreman-Peck 1988: 257–78.

Galambos, L. 1988. "Looking for the Boundaries of Technological Determinism: A Brief History of the U.S. Telephone System." In Mayntz and Hughes 1988.

Gebhardt, H.-P. 1989. "Rechtsgrundlagen des Datenschutzes sowie Datenschutz im Fernmeldewesen der Länder Schweiz, Frankreich, Niederlande, Großbritannien, Schweden, USA und Japan." In *Archiv für das Post- und Fernmeldewesen* 2: 163–69.

Gibbons, J. H., and Gwin, H. L. 1985. "Technology and Governance." *Technology In Society* 7: 333–52.

Grande, E. 1989. *Vom Monopol zum Wettbewerb? Die neokonservative Reform der Telekommunikation in Großbritannien und der Bundesrepublik Deutschland.* Wiesbaden: Dt. Univ. Verlag.

Granstrand, O., and Sigurdson, J. 1985. "The Role of Public Procurement in Technological Innovation and Industrial Development in the Telecommunication Sector. The Case of Sweden." In *Technological Innovation and Industrial Development in Telecommunications*, ed. O. Granstrand and J. Sigurdson, pp. 149–72. Gothenburg.

Harris, R. G. 1989. "Telecommunications Policy in Japan: Lessons for the U.S." In *California Management Review* 3: 113–31.

Heuermann, A.; Neumann, K.-H.; and Schnöring, Th. 1986. *Telekommunikationspolitik im Vergleich—Eine problemorientierte Übersicht über die Länder USA, Großbritannien, Frankreich, Niederlande, Schweden und Japan.* Bad Windsheim: Heidecker.

Hills, J. 1989. "Neo-conservative Regimes and Convergence in Telecommunications Policy." In *European Journal of Political Research* 17: 95–113.

Holecombe, A. 1911. *Public Ownership of Telephones on the Continent of Europe.* Cambridge: Harvard University Press.

Hollingsworth, J. R., and Lindberg, L. N. 1985. "The Governance of the American Economy: The Role of Markets, Clans, Hierarchies, and Associative Behaviour." In *Private Interest Government*, ed. W. Streeck and P. C. Schmitter. London and Beverly Hills: Sage.

Hughes, T. P. 1987. "The Evolution of Large Technical Systems." In W. Bijker, T. Hughes, and Pinch, T., *The Social Construction of Technological Systems.* (Cambridge: MIT Press).

Hyden, G. 1988. "Dilemmas of Governance in Africa: Report on Work in Progress." Paper presented at the XIVth IPSA World Congress, Washington, D.C., August/September 1988.

Kaijser, A. 1987. "From Local Networks to National Systems. A Comparison of the Emergence of Electricity and Telephony in Sweden." In *1880–1980: Un Siècle D'Électricité dans le Monde*, ed. F. Cardot, pp. 7–22. Paris: Presses Universitaires de France.

Katz, M. L., and Shapiro, C. 1985. "Network Externalities, Competition, and Compatibility." *American Economic Review* 75: 424–40.

Katzenstein, P. 1987. *Policy and Politics in West Germany*. Philadelphia: Temple University Press.

La Porte, Todd R. 1984. *Technology as Social Organization*. IGS Studies in Public Organization Working Paper No. 84-1. Berkeley: Institute of Governmental Studies, University of California.

Lehmbruch, G.; Singer, O.; Grande, E.; and Döhler, M. 1988. "Institutionelle Bedingungen ordnungspolitischen Strategiewechsels im internationalen Vergleich." In *Staatstätigkeit. International und historisch vergleichende Analysen* (PVS Sonderheft 19), ed. M. G. Schmidt. Opladen: Westdeutscher Verlag.

Libois, L.-J. 1983. *Genèse et Croissance des Télécommunication*. Paris: Masson.

Lindberg, L. N., Campbell, J. L.; and Hollingsworth, J. R. Forthcoming. "Economic Governance and the Analysis of Structural Change in the American Economy." In *The Governance of American Economy*, ed. J. L. Campbell, J. R. Hollingsworth, and L. N. Lindberg, pp. 1–82.

March, J. G., and Olsen, J. P. 1984. "The New Institutionalism: Organizational Factors in Political Life." *American Political Science Review* 78: 734–48.

Mayntz, R. 1988. "Zur Entwicklung technischer Infrastruktursysteme." In *Differenzierung und Verselbständigung. Zur Entwicklung gesellschaftlicher Teilsysteme*, ed. R. Mayntz, B. Rosewitz, U. Schimanck, and R. Stichweh, pp. 233–59. Frankfurt: Campus.

Mayntz, R., and Hughes, T. P., eds. 1988. *The Development of Large Technical Systems*. Frankfurt: Campus; Boulder: Westview.

Metelski, J. 1978. "Telecommunications Privacy and the Information Society." *Telecommunications Policy* 4: 327–35.

Mitchell, B. M. 1983. "The Cost of Telephone Service. An International Comparison of Rates in Major Countries." *Telecommunications Policy* 2: 53–63.

Müller, J., and Foreman-Peck, J., eds. 1988. *European Telecommunications Organisation*. Baden-Baden: Nomos.

Neumann, K.-H. 1987. *Die Neuorganisation der Telekommunikation in Japan*. Berlin: Springer.

Noam, E. M. 1989. "International Telecommunications in Transition." In *Changing the Rules: Technological Change, International Competition, and Regulation in Communications*, ed. R. W. Crandall and K. Flamm, pp. 257–97. Washington, D.C.: Brookings Institution.

North, D. C. 1981. *Structure and Change in Economic History*. New York and London: Norton.

OECD 1973. *Computers and Telecommunications*. Paris: OECD.

———. 1987. *Trends of Change in Telecommunications Policy*. Paris: OECD.

———. 1989. "Present Situation and Trends in Privacy Protection in the OECD Area. *STI Review* 6 (December): 179–213.

Ouchi, W. G. 1980. "Markets, Bureaucracies and Clans." *Administrative Science Quarterly* 25: 129–45.

Perrow, C. 1984. *Normal Accidents: Living with the High-Risk Technologies*. New York: Basic Books.

Schmitter, P. C. 1989. "Sectors in Modern Capitalism: Modes of Governance and Variations in Performance" ["*I settori nel capitalismo moderno: modi di regolazione e variazione nel rendimento*"]. *Stato e mercato* 26 (1989): 173–208.

Schneider, V. 1989. *Technikentwicklung zwischen Politik und Markt: Der Fall Bildschirmtext*. Frankfurt: Campus.

Schneider, V.; Vedel, T.; and J. Miller. 1990. "The State and New Forms of Communication: The Introduction of Videotex in Europe and North America." In *State Policies and Techno-Industrial Innovation*, ed. Ulrich Hilpert. London: Routledge.

Shepsle, K. A. 1989. "Studying Institutions. Some Lessons from the Rational Choice Approach." *Journal of Theoretical Politics* 2: 131–47.

Snow, M. S. 1982. "Telecommunications and Media Policy in West Germany: Recent Developments." *Journal of Communication* 2: 10–32.

Takanashi, N., et al. 1988. "The Achilles' Heel of the Information Society: Socioeconomic Impacts of the Telecommunication Cable Fire in the Setagaya Telephone Office, Tokyo." *Technological Forecasting and Social Change* 34: 27–52.

Thomas, F. 1988. "The Politics of Growth: The German Telephone System." In Mayntz and Hughes 1988, pp. 179–209.

Tomita, T. 1984: "Japan's Policy on Monopoly and Competition in Telecommunications." *Telecommunications Policy* 8, 1: 44–50.

Vedel, T. 1988. "La Déréglementation des Télécommunications en France: Politique et Jeu Politique." In *Le Déréglementations. Étude Comparative*, ed. B. Chend, pp. 281–312. Paris: Economica.

Weingart, P. 1989. "Großtechnische Systeme: ein Paradigma der Verknüpfung von Technikentwicklung und sozialem Wandel." In *Technik als sozialer Prozeß*, ed. P. Weingart, pp. 174–96. Frankfurt: Suhrkamp.

Werle, R. 1990. *Telekommunikation in der Bundesrepublik: Expansion, Differenzierung, Transformation*. Frankfurt: Campus.

Wieland, B. 1985. *Die Entflechtung des amerikanischen Fernmeldemonopols*. Berlin: Springer.

Wieland, B. 1988. "Telecommunications in Netherlands." In Müller and Foreman-Peck 1988, pp. 203-219.

Williamson, O. E. 1985. "Reflections on the Institutional Economics." *Zeitschrift für die gesamte Staatswissenschaft* 141: 187–95.

Williamson, E. O. 1985. *The Economic Institutions of Capitalism*. New York: Free Press.

——————. 1986. "The Economics of Governance: Framework and Implications." In *Economics as a Process. Essays in the New Institutional Economics*, ed. R. N. Langlois. Cambridge: Cambridge University Press.

Wittfogel, K. A. 1962. *Die orientalische Despotie. Eine vergleichende Untersuchung totaler Macht*. Köln: Kiepenheuer & Witsch.

THE EXTERNAL CONTROL OF LARGE TECHNICAL SYSTEMS: OFFSHORE OIL PRODUCTION SAFETY AND HEALTH REGULATION IN GREAT BRITAIN AND NORWAY

Svein S. Andersen
Norwegian School of Management

The External Regulation of Large Technical Systems

This article deals with the external regulation of a large technical system (LTS), the offshore contractor-operated systems of oil production in the North Sea. These systems are often thought of as infrastructures providing input for or carrying important social processes (electricity grids, communication systems) integrated across geographical space. Here we look at the legal and organizational structure of the offshore oil production system as an infrastructure for state strategies of safety and health regulation. One of its properties is that offshore oil activities violate basic assumptions of traditional state regulation in Western Europe.

We can distinguish three degrees of regulation by how ambitious their regulatory goals are:

1) To ensure minimum legal requirements, such as the demand that national subsidiaries should be registered in host countries.
2) To make sure that company activities follow certain rules or technical standards.
3) To ensure the quality of system design and management (e.g., by reducing operational and occupational risk) so that certain outcomes do not occur.

Traditionally, safety and health regulation has been formulated in terms of rules and standards that have been externally imposed. This is consistent with formal-legal bureaucratic control of liberal societies. However, in some countries goals of safety and health regulation have been formulated in more general terms, to be implemented through arrangements of social cooperation. In these cases strict monitoring and sanctioning power have been supplemented with corporatist arrangements.

The nature of LTSs makes external regulation of safety and health particularly difficult, since a) regulations and standards relate to specific processes or issues rather than the whole system at hand; b) even if external experts are well qualified, they lack the "hands on" experience to make independent judgments; and c) state intervention may be viewed as undue interference by companies controlling the access to the system. In such situations, it is difficult for regulatory agencies to assure that particular goals and standards will be achieved through traditional strategies. A different approach is needed.

The contractor-operated oil production systems are a complex combination of technologies, organization, and personnel from many sectors. In the North Sea system, they

T. R. La Porte (ed.), Social Responses to Large Technical Systems, 43–59.

are linked in new ways under the authority of the oil companies with operator responsibility (see figure 1). There is considerable variation among them as their technical and managerial expertise have been continually elaborated to fit the conditions of the North Sea. But one common characteristic is that they have vehemently resisted external regulation, using their control over physical access to offshore installations as a key defensive resource.

After a brief overview of the Norwegian and British cases, this article shows how both national models of regulations of safety and health have failed in the North Sea. In response, a new regulatory model has been developed in the Norwegian offshore sector and is now being introduced by Norwegian authorities as a general model of regulation in a complex modern society. The new model has three main elements:

1) Regulatory authorities no longer attempt to specify goals or standards to be fulfilled. The overriding object is to commit operators to work for continued system improvement.
2) The authority of the regulatory relation rests on "rational consensus" between public and private experts, as a basis for procedural rationality to deal with the intrinsic technical uncertainty of the unique system to be regulated.
3) The regulatory system incorporates elements of corporatist participation to secure employee safety. (However, this has been less important than the technical risk.)

Models of Regulation

Oil production started on the Ekofisk field on the Norwegian shelf in 1971, and off Great Britain in 1975. From the mid-1970s, offshore activities developed rapidly (see figure 2). We concentrate on organizational problems of safety and health control and on attempts to improve the situation since the mid-1970s.

There are three main groups of companies operating in the North Sea—the multinational oil companies, the international service and supply industry, and the national service and supply companies. We are particularly interested in the multinational oil companies with operator responsibilities. They have been responsible for the development of the offshore oil systems, and the most reluctant to accept state regulation. The operators have, since the mid-1960s, actively challenged state regulation of safety and health in the North Sea. Compared to onshore industries, offshore safety and work environment performance has been poor in both countries, from three to eleven times more accidents than that of onshore industrial sectors (Andersen 1988: 71). The companies have used fairly similar standard stratagems to limit government intervention, exploiting ownership prerogatives and their control over isolated offshore installations.

Even though the nature of the problems facing regulatory authorities in the two countries has been similar, responses have been very different. Neither government

FIGURE 1

Typical Pattern of Project Organization and Allocation of Management Responsibilities in the North Sea

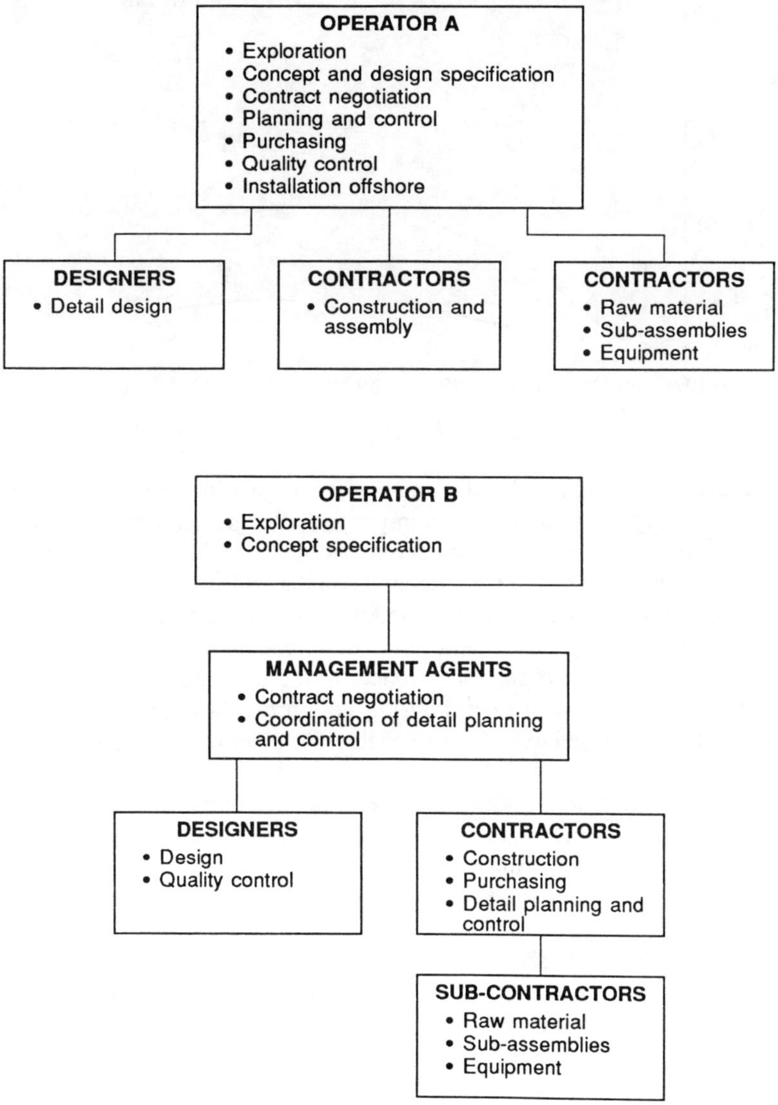

Source: *North Sea Costs Escalation Study*, Part II, p. 97.

FIGURE 2

Offshore Production in Great Britain and Norway

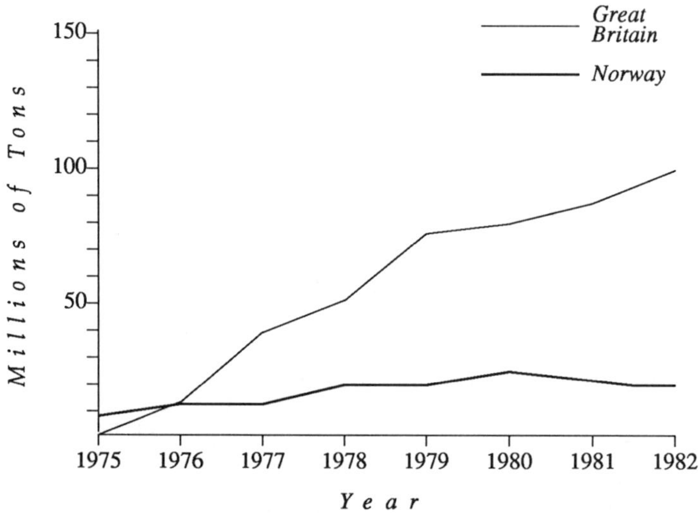

intervened directly in offshore health and safety regulation. Responsible regulatory authorities were mainly left to handle oil company challenges on their own. An important element in achieving effective regulation stems from the ability to reduce technical and administrative complexity. This, in turn, depends on the ability of legal and administrative traditions to define new issues in a coherent way, on the authorities' ability and willingness to learn from experience and to handle the technical problems of implementing regulation. In this respect regulatory models in the two countries vary considerably. Although Britain and Norway are neighboring countries, they represent different traditions of safety and health regulation—those of the liberal and the neo-corporatist systems, respectively.

In Britain, the legal and administrative system has traditionally been characterized by sectoral specialization and piecemeal elaboration of legislation. Legal rules in different sectors may overlap and there may be considerable competition between regulatory agencies. Each agency develops independent expertise. The adversary nature of regulatory relations is modified by the assumption that the private companies are fellow players (Wall 1976). However, regulation is based on relatively active sanctioning and control.

In Norway, the legal system is unitary and the administrative responsibility is handled by one agency, although authority may be delegated. Safety and health regulation is independent of sector authorities. The Norwegian "soft-law" tradition views the regulatory relation as based on a fundamentally cooperative attitude on the parts of those who are controlled. Affected parties are assumed to share the objectives and intentions laid down in rules and regulations, partly because they participate in the making of such rules. Less emphasis is put on strict monitoring and sanctioning, although the formal basis for the exercise of state authority does exist.

Neither the British nor the Norwegian system was adequate to the challenges of the North Sea. British authorities have consistently been unable to make traditional legal-bureaucratic regulation work effectively. There have been serious problems of legislation in the organization of control activities and in their implementation. Despite a general reform toward a unitary regulatory system in the mid-1970s, the responsibility for offshore oil safety and health remained in the sector. Legal and interagency tangles created delays. There was a high degree of personnel exchange between regulatory authorities and the oil companies. Regulatory experts tended to look at themselves as "oil industry" men. Offshore unions were too weak to have any influence, even to serve as a source of alternative information.

In Norway, the authorities did not even attempt to apply the "soft-law" model, which had emerged as part of a general corporatist development. In this model, general rules and regulations are supported by tripartite arrangements, and implementation rests on the tacit cooperation of the parties involved. The authorities felt that foreign oil companies could not be trusted to respect the general goals and values of Norwegian society. Therefore, the Norwegian authorities fell back on a strict legal-bureaucratic system, keeping the foreign oil companies at arm's length during the preparatory stages. Offshore unions were brought in as important allies. This simplified legislation and the creation of a new regulatory agency. However, it was soon apparent that this would not work in the implementation stage. The authorities had no expertise of their own, and they felt that experts in the industry could not be trusted. In addition, the companies used their control and physical access to offshore installations as a means of reducing state interference.

In contrast to Great Britain, Norwegian authorities have gradually evolved a new form of regulation, developed by the Norwegian Petroleum Directorate (NPD). It has three main elements:

1) The authorities do not attempt to formulate standards or specific goals for company activities in specific areas. Although formal authority is not deferred, the companies have the day-to-day responsibility for continued system improvement.
2) The core of the regulatory system is a body of procedures for the handling of safety and health issues in a holistic way. The content of such rules is arrived at through discourse between public and private expertise. Cooperation is not based on social consensus, but rather on "rational consensus."
3) While spot controls do exist, they are not an important part of the system. The external control concentrates on the system of rules governing safety and health in the company. Rather than a traditional authority relation, this is a system of self help. The role of the state is to organize the process of rule-making and system development to rationalize offshore systems.

As part of a general reform toward tripartite control systems, offshore unions were brought in through local safety and health committees. However, the working of such a system was complicated by the fact that the unions in the operator companies were not part of the Norwegian corporatist system. To further complicate matters, existing legislation was not adequate to cover the complex, large-scale, and strict hierarchic organization found on offshore platforms.

We now describe the development of British and Norwegian offshore safety and health regulation in greater detail, in terms of rule-making and the creation of an administrative system and its implementation. We conclude with a discussion of different forms of regulation in relation to LTSs.

Legislative Framework

The offshore oil sector was a new "multi-sector" industry. In such sectors the legislative process faces two important problems: first, to what extent does the legal framework cover the whole system in question; and second, to what extent are authorities able to preserve a sufficient autonomy from those regulated, to specify the public interest independent of company concerns.

Until the early 1960s, the oil industry operated in a legal vacuum offshore. The first offshore legislation in Norway (1963) and the United Kingdom (1964) established general frameworks for state authority over extraction of natural resources on the continental shelf. At the time, neither country gave much attention to safety and work environment issues. The implicit notion was that onshore safety regulations could be extended offshore. As it turned out, this issue was not so straightforward. Offshore activities necessitated special regulation beyond the factory and shipping acts. These requirements created overlap, grey zones, and incompatibility in relation to the established system, particularly in Great Britain.

For a long time, British efforts to set up a legal framework for offshore operations developed very slowly. But in 1963, the government responded to company initiatives for general legislation with what has been termed an "unseemly haste." With doubts about the momentum underway, concern for the adequacy of safety arrangements was not allowed to impede the rapid issuing of provisions. For this reason, the implications of the 1964 Act for Occupational Safety Offshore were not clear. One interpretation was that relevant sections of the 1934 Petroleum (Production) Act covering onshore activities should be extended offshore. Another was that parts of the Factories Act of 1961 should be made appropriate.

To clarify the situation, an additional sub-section was inserted into the clause dealing with licensing. It stated that model clauses prescribed under the 1934 Petroleum Act should include provisions for safety, health, and welfare of persons employed under the authority of any licensee (Carson 1982: 147). This meant that the licensees were required to comply with any instruction given by the minister. This system relied on the given code for the substance of provisions and the contractual nature of the license for its enforcement. The only efficient sanction was to revoke the contract.

The shortcomings of the British system were demonstrated with the collapse of the "jack-up" rig *Sea Gem* in 1965, in which thirteen lives were lost. The committee investigating the accident concluded that state authorities lacked proper control. New legislation was proposed, although the intention to do so was not announced until 1969. Finally, the Mineral Working (Offshore) Bill was passed in 1971, empowering the Secretary of State to make regulations. It was to become the main legal instrument for controlling offshore safety organization in the years to come; it did not, however, immediately become operational.

The 1971 Act was tailor made to give the offshore sector its own safety legislation. It was passed in the same year that the Robens Committee published its first report preparing a general reform of the British safety control system. This report concluded that one of the main deficiencies of the existing system was the type of piecemeal provisions proposed for the offshore sector. Up to 1973 the regulations passed dealt mainly with issues of administrative infrastructure. The first rules concerning the substance of regulation came the following year, and a number of others were passed in the years to come.

In Norway, offshore safety legislation was less problematic than in Great Britain. Within the framework of the 1963 Continental Shelf Act, the State Oil Council proposed a set of special safety regulations. They covered offshore exploration and drilling and became operational in 1967. It was not clear whether the Workers' Protection Act of 1956 applied offshore. Among other things, this law placed responsibility for safety on employers, and allowed for workers' representatives to participate in safety control. However, in the first years all mobile rigs were foreign-owned and not covered by Norwegian law.

The 1967 regulations were also adapted to the production activities which started on Ekofisk in 1970. This extension was based on a special agreement between the Ministry of Industry and Phillips Company. However, regulations did not cover working hours, local safety committees, or job security. This created uncertainty within the administration about authority rights and delegation of responsibilities. Such issues were accentuated by a serious fire on Ekofisk in 1975, in which five men died. This event made offshore safety a major issue on the political agenda (Graver 1983: 66). Soon after the accident the Halden Committee was created. It included offshore unions as well as management and government representatives.

The committee's work resulted in two major amendments. In 1976, general work-protection regulations were extended to production installations. This meant that important parts of the Workers' Protection Act of 1956 were applied. In maritime law such issues were covered by several acts, most importantly the Seamen's Act of 1975. The Ministry of Trade (and Shipping) created regulations applying to all floating installations from 1 January 1977, and in February 1977 a new general Work Environment Act was passed. In June of the same year, the special offshore regulations were temporarily added to the Work Environment Act. A year later, they were made permanent.

The 1977 blow-out on Ekofisk Bravo focused attention on aspects of operational safety. However, the 1976 and 1977 work environment reform represented the culmination of political reforms. Further developments have been linked to the administrative development, particularly the role of the Petroleum Directorate. We will return to this in the next section.

To summarize:

> The creation of the British legal regime took time. The assumption was that the oil sector was special and needed a safety regime of its own. Economic and technical uncertainty implied that rules should not be too detailed. Consequently, the legislation followed the basic logic of traditional regulation without being very specific. This reflected the wishes of the oil industry.
>
> The Norwegian regulatory system developed more quickly. The oil industry was held at arm's length. Existing general maritime and onshore safety regimes were extended with some modifications. Offshore unions (at the time, only the Seamen's Union and the Operator Association on the Ekofisk field) were actively involved. The approach was holistic, in the sense that there was little conflict between sectoral bodies. Basically, the same laws were valid for all fixed and floating installations. However, there was no concept of offshore *systems* involved in the legislative framework.

The next step was to develop administrative apparatuses to implement regulations.

Administrative Organization and Regulation

The legislation stated general goals and delegated responsibilities for the making of further regulations to administrative agencies. The problem was how to operationalize such general mandates into specific solutions and routines.

In Great Britain, underlying political controversies were reflected in interagency conflicts. Technical experts from the oil industry were employed. In Norway, the formation of an administrative apparatus for offshore safety was complicated by two other factors. First, it was paralleled by a general reform of work environment control where local safety committees were to play a major role; second, time was needed to develop independent in-house expertise.

In Britain the policy responsibility for offshore safety moved between a number of institutions due to reorganizations motivated by other concerns. In the 1960s the Petroleum Division was transferred from the Ministry of Power to the Ministry of Technology. In 1970, it went to the Department of Trade and Industry. In 1973, there were two divisions dealing with oil: the Oil Policy Division (two branches) and the Petroleum Production Division (three branches). With the exception of pipelines, safety was primarily the responsibility of the Petroleum Production Division. In 1977, the Petroleum Engineering Division (PED) took over most operational and safety matters.

Despite all these changes, there is a major consistency in the handling of offshore safety issues. They were firmly located within the sector administration which also was responsible for such matters as licensing and general North Sea policy. Within the different departments, officials responsible for safety were answerable to superiors whose brief included other, and not necessarily compatible, objectives (Carson 1982: 165). The "sponsoring department" looking after safety was the normal pattern in British industry up to 1974. After the general reorganization of safety inspection in the mid-1970s, this arrangement persisted only in the petroleum sector.

The offshore sector involved a multiplicity of activities. A number of preexisting agencies became responsible for particular operations. There were often disagreements over demarcation lines. For instance, the Department of Trade's Marine Division and Civil Aviation Authority's Marine Division were under the Merchant Shipping Act responsible for safety on vessels and diving operations on British ships. The first agency had also played an important role in drafting regulations pertaining to life-saving appliances and fire-fighting equipment. Its surveyors continued to carry out the relevant examination and certification on behalf of PED.

A major conflict developed, not surprisingly, between PED and the new Health and Safety Commission (HSC) and its executive branch (HSE). Since 1974, the latter agencies were responsible for the general and unified control system covering all other industries. In 1975, the HSC attempted to take over occupational safety on offshore installations. Structural safety and blow-out risk should remain in PED. An agreement was reached whereby PED would inspect offshore installations on HSC's behalf. HSE was only to assume direct responsibility for occupational safety in connection with pipe-laying operations and crane ships.

The implementation of the agency agreement was delayed until November 1978. New forms of offshore control based on employee participation were complicated by weak unions. The same year the Borgoyne Committee was formed to evaluate the conflict between PED and the oil industry, on the one hand, and HSC/HSE, on the other. The committee reported in 1980; in the meantime, the Labour party government had been replaced by a Conservative one. The committee's conclusion was that there was a need for a unified and specialized authority on safety matters in the sector. PED got back the full responsibility.

In Norway as well, several agencies became involved in the implementation of offshore safety and work environment regulations. However, this did not lead to intense struggles over administrative responsibilities. The main responsibility was shared by two institutions with the assistance of a number of others. The Petroleum Directorate (NPD) directly inspects or coordinates control for all production-related activities. The Norwegian Maritime Directorate takes care of all ships and flotation installations. This corresponds to the division of legal responsibilities between the Work Environment Act (1977) and the Seamen's Act (1975).

The NPD was formed 1 April 1973. It took some time to build up organization and competence. In its first years, the agency had a technical orientation; the issue of work protection was first mentioned in a special section of the NPD yearly report of 1976. Up to 1979, the agency reported to the sector administration, the Ministry of Petroleum and Energy (previously the Ministry of Industry). From January 1979, the political responsibility for offshore safety was transferred to the Ministry of Local Government and Labour. This ministry has the general responsibility for safety issues, according to a change initiated by the Norwegian Federation of Trade Unions.

The serious accident on Ekofisk in 1975 affected legislation; the administrative organization was influenced by the Bravo blow-out in 1977 and the capsizing of the Alexander Kielland in 1980. In both cases, existing control systems were evaluated. The main organization remained the same, but reporting routines for the companies were changed. More time was put into preventive work at the design and planning stage. Still, there was a general bias toward technical issues rather than work organization. This tendency also characterized further legislation.

In the Halden Committee, the operators were treated as outsiders. In contrast to the unions, they were excluded from the reform work and only consulted at a late stage. As the North Sea Operator Committee pointed out in their comments, this was a "regrettable" departure from Norwegian practices. However, in subsequent regulatory legislation the oil companies have played an important role. As operators they control important information about technology as well as offshore conditions that the authorities need. This is demonstrated in the NPD's "hearing" procedures.

The hearing is a central element in the devising of new rules and statutes, and is an integrated part of Norwegian administrative practice laid down in the law. The NPD procedure differs from the normal pattern in several ways. For one thing, it is more formal and bureaucratized. There is a clear distinction between the directorate's preparations and a hearing. However, while it is common only to include interest organizations, NPD provides single oil companies with an important role. This arrangement reflects both the position and the size of the companies in the Norwegian context (Graver 1983: 166–77).

The impact of this system must be viewed in connection with two other aspects of the Norwegian offshore control system. According to the Work Environment Act of 1977, employers are responsible for their internal control. Each company has to develop safety rules, and to work toward continuous improvements. Safety is part of the central "line" responsibility. Workers are given the right to participate in local safety committees. State agencies have a network of local representatives. They concentrate on control of internal safety rules and act as a forum for appeal when local parties cannot reach an agreement.

These requirements of the new Work Environment Act represented a radical reform in the Norwegian onshore industry. It extended tripartite arrangements to the domain of safety and work-environment control. Onshore the main problem had been to clarify the authority and negotiation rights given to employers and worker representatives. Offshore operators and contractors at first neglected development of safety regulations and obstructed the formation of safety committees. After 1980, the situation improved among operators and larger contractors, although inter-union conflicts still created problems (Karlsen 1982). Nevertheless, such arrangements had important implications for the content of regulations when they were implemented. We will return to this below.

To summarize:

> The British system was characterized by inter-departmental conflicts around administrative responsibilities and control strategies. The political coalitions

were clearly defined. The PED was supported by the oil industry. The HSC/HSE and their allies included the union representatives. This political-administrative organization of offshore responsibilities added to the intrinsic complexity of the sector instead of reducing it.

In Norway, administrative responsibilities were divided on the basis of the preexisting distinction between maritime and onshore activities. Problems were related to lack of administrative organization and competence. The transfer of political responsibility to the ministry with general responsibility for safety did not create conflict. It was an orderly legal-bureaucratic system.

Implementation

Traditionally implementation is supposed to be the application of preestablished rules, administrative techniques, and standardized solutions to a well-defined set of problems. The offshore situation, however, was ambiguous. In addition, rules and an administrative organization took time to develop. Attempts at control paralleled the evolution of control systems. In Britain, problems of implementation entailed difficulties in finding legal-technical and administrative solutions, but shortcomings were tolerated. In Norway, fundamental preconditions for cooperative implementation were clearly not fulfilled. This led to a reevaluation of the traditional control model.

In the early days of offshore activities, inspectors in the British PED not only had to administer rules, they also had to make them. Many inspectors were involved in writing the regulations, even as they were attempting to enforce them. They were specialists on narrow technical issues rather than on administrative control and came to regard regulations in a special light. Regulations were not infallible rules which had to be obeyed, but the product of fallible human beings. In addition, making rules was time-consuming. Too little time was left for inspection.

During the formative years, the number of inspectors was small, the work load was high, and there was a traffic in personnel between the regulatory agency and the regulated. A special relation developed based on the shared assumption that offshore activities were unique. A key factor in this was the technology. The industry expected inspectors to be drawn from their own ranks. Some, like union representatives, argued that PED acted as the industry's advocates rather than its adversaries (Carson 1982: 176). This criticism was dismissed by the Borgoyne Committee in 1980.

British offshore safety control has been characterized by deference to the industry's technological status. An important constraint has been to avoid steps which could hamper technological progress. Attempts to impose detailed safety control were often highly contentious. As a representative of ESSO put it, "Requirements should be put into guidance rules, to clarify the intentions of legislation, but not bind an operator to any particular course of action" (Carson 1982: 179). Legal rules had a tendency to become technologically negotiable.

For inspection and investigation of accidents on installations inspectors have had to rely on the companies for helicopter transport. In the cases of installations, as opposed to barges, PED inspectors have since 1973 had the right to require the provision of transport at "any reasonable time." About one week's notice is common. This means that spot checks have been precluded, since managers will have time to take at least some remedial action in advance. Inspections will be limited to what falls within the specialized (and narrow) competence of the inspector (Carson 1982: 239–40).

Finding procedures of enforcement is usually a matter of trial and error. In the case of British offshore safety this learning process was complicated by a number of factors. Existing laws and administrative organization had a series of commissions, unclear definitions and disarticulation between segments of the bureaucracy. Such anomalies and loopholes were effectively exploited by the companies.

Implementation has also been complicated by the fact that when offshore activity moved north, parts of it fell under the jurisdiction of Scottish law. The general offshore control system was organized from the center. Its implementation would eventually depend on the legal and administrative system prevailing in the "periphery." The decision to prosecute, for instance, rested with the Scottish Procurator Fiscal. The mismatch between central and local levels also complicated the investigation of accidents (Carson 1982: 264–65).

Even when authorities were sufficiently confident to pursue legal sanctions, their success was by no means guaranteed. Five of the thirteen cases which reached the courts between 1977 and 1980 resulted in "no finding of any guilt against any party." Two out of three defendants in a further case were found not guilty. Of the twenty-three companies or individuals involved only ten were admonished. The fines varied between 50 and 400 pounds—hardly enough to defer any company (Carson 1982: 267).

The implementation of safety and work-environment control in Norway differed from the British experience in a number of ways. There was less confusion over legislation, and regulation developed faster. There were no significant administrative conflicts. From the late 1970s, the legal basis, regulations, and administrative responsibilities covering all offshore activities were harmonized. Until the 1970s, problems of implementation reflected lack of resources for strict monitoring and sanctioning. Later reforms focused on employers "internal control" and active participation from local safety delegates and committees.

The British offshore system was based on a 150-year-old tradition of specialized legal-bureaucratic control. In the Norwegian tradition, implementation was less strict and based more upon compliance. This difference has been more pronounced since the 1950s, and with the general reform in 1977 central control was to be increasingly indirect and general. In correspondence with this, Norwegian offshore regulations have been characterized by two tendencies. First, general goals tend to become more specific while fewer rules are given in the form of regulations. Second, regulation is based on considerable experience and lasting negotiations (Graver 1983: 184).

The development of "internal control" became the solution to the implementation problem. In this system, the employer is responsible for safety as part of line management. This "self control," as it was called in the first years, is to be supervised by the relevant state agencies. On multi-employer work sites, like productions platforms, operators are responsible for all employees. The control system combines delegation with a strengthening of hierarchic control over the development of local safety regimes. This control should ensure that all activities are planned, organized, implemented, and maintained in correspondence with standards given in laws and regulations.

Licensees should be able to demonstrate that their control system is good enough. In 1979, the NPD provided a 125-point guideline about the most important parts that had to be covered. Experts from the directorate supervise internal control systems through "professional discussions" with company representatives. Since the companies are assumed to share an interest in maximum safety questions are expected to be answered "openly and honestly." A problem has been that up to the late 1970s, NPD's competence was clearly behind that of the companies'—good (but "flexible") public experts could make interesting and well-paid careers in the companies. During the last few years the situation has improved.

This system of internal control has been evaluated in relation to the major offshore accidents. The conclusion was that it was satisfactory, but it should be made more efficient. However, the focus has been on structural safety. The question of safety in relation to work organization has attracted less attention. This issue was closely related to the system of employee participation.

Local safety delegates are to be elected by employees; sometimes also by the union office. They are to participate in local control systems. The Work Environment Act allows for many representatives. The operator companies have to allow safety committees among all operators and the largest contractors. However, safety work has been fragmented, despite regulations providing a good basis for joint committees. A major obstacle has been the conflict between unions.

The intention is that individual employees shall be able to find solutions to problems which they face, and most issues are solved this way. When this cannot be done, problems have to be handled through the local safety system. The Work Environment Act does not specify how such arrangements are supposed to be organized. In practice, it tends to become a "shadow organization" of the company. On the offshore installations, safety systems reflect the extremely hierarchic and specialized work organization of operators and large contractors (Karlsen 1982: 177–78).

Figure 3 illustrates problem-solving related to offshore safety and work environment. The left side shows arrangements found in the operator company, the right side represents the system of the contractor company (usually more than one). Each of the main parts is divided in two—safety systems and the technical and economic organization. Channels for reporting rapidly become complex and time-consuming to pursue. Karlsen reports on a procedure with fourteen levels—more than NATO headquarters in Central Europe uses to contact a frontier soldier in northern Norway (Karlsen 1982: 187).

The local safety organization arrangement has prompted several developments. Safety delegates have simplified the system by establishing a direct hotline to NPD. This creates an overload on the central level. And the complex organizational context stimulates strategic games and formal regulation (Karlsen 1982: 160–72). In the last few years the

FIGURE 3

Problem Solving Concerning Work Environment Protection in Norway

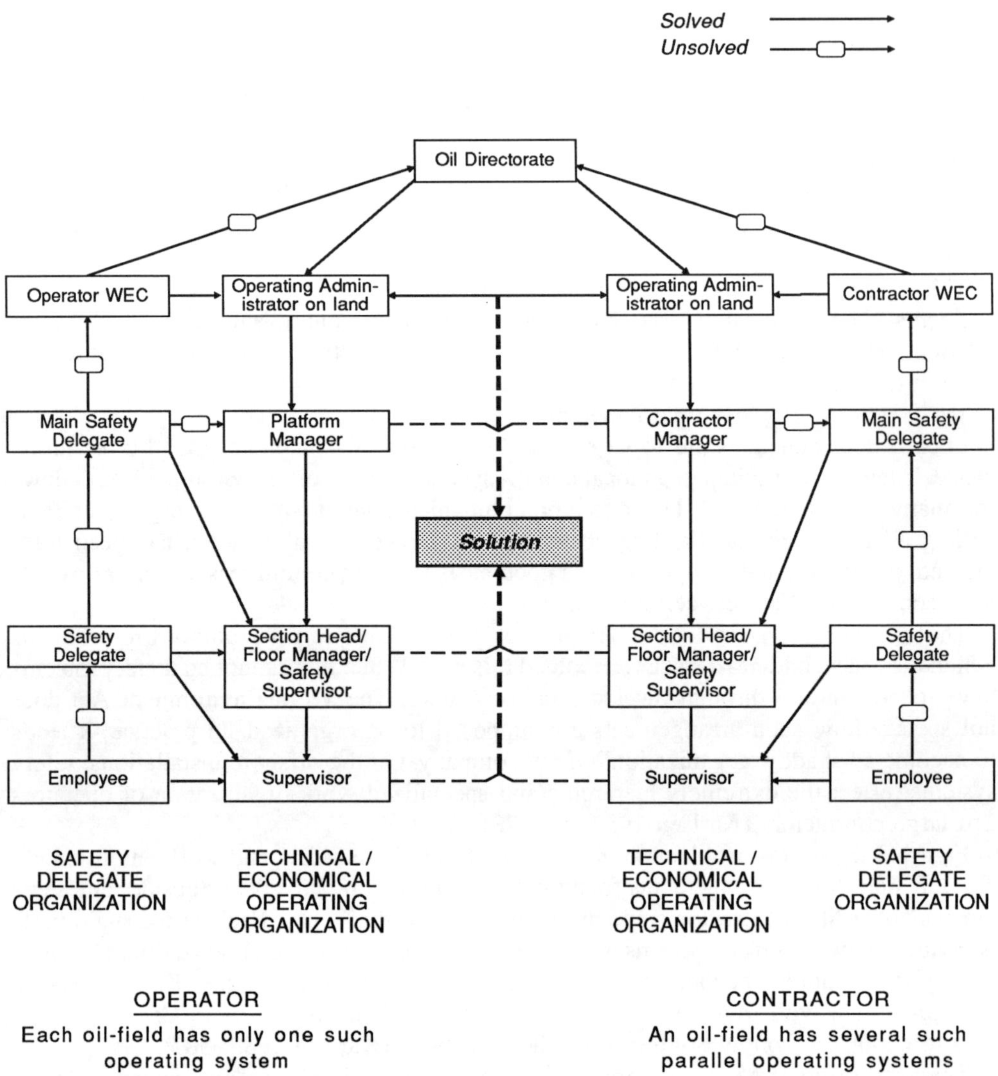

Source: Karlesen, 1982: 179.

system seems to have worked reasonably well. A large majority of employee representatives thinks it is a suitable system. Also, the most radical clauses in the Work Environment Act are exercised, i.e., the right to stop hazardous work—some 49 percent of the safety delegates said they had done so (Karlsen 1982: 143–55). They are, however, more efficient in solving everyday problems than in raising general and long-term issues (Liaaen 1982: 33).

To summarize:

> It is clear that offshore regulation in Britain and Norway is not hierarchic in any strict sense. In both countries, state agencies provide information and support to the industry. They try to stimulate learning and research related to accidents. There is an element of cooperation and negotiation in both systems.
>
> In Britain, such relations have been internalized in the central administrative structure as informal arrangements based on political alliances and exchange of expertise. Few conflicts are expressed in relation to implementation of rules, despite the widespread failure of the system.
>
> In Norway, cooperation at the implementation stage is built around a two-tier system: the core of the system has been bipartite between the NPD and the operators related to technical issues. Tripartite structures aimed at safety control related to internal control have not functioned as intended.

LTSs and Forms of Regulation

These LTSs put special demands on the regulation of safety and health. As this article shows, traditional British and Norwegian models of regulations have failed to meet these demands in the North Sea.

Precisely because Norwegian authorities felt that the "soft-law" tradition was vulnerable, attention was given to the social and organizational limits of regulatory control. Still, it took time before a new system could be developed. Corporatist elements were used to form a bipartite alliance between a new, weak agency and offshore unions. The most important regulatory measure was the development of the "internal control" system.

The new system demonstrates how one can approach the problem of regulating LTSs not by imposing rules and standards, but through procedures that can improve the quality of both organization and management.

Efficient regulation has to take into account that LTSs are unique systems. At the same time, normative models of organization and management are not so developed that they can serve as a basis for detailed guidelines. Some concrete demands can be formulated on the basis of knowledge of what may go wrong. However, in the end the challenge is to create a system bridging external control ambitions and the interests of those to be regulated. In systems with a strong technical core, where the consequences of accidents can be very costly, there is a basis for a common course.

It is not uncommon for state authorities to base their regulation on the activities of private actors, in areas where there is no transfer of formal authority. In the "negotiating society" (Olsen 1982) public decisions are to a large extent based on information and expertise from private parties. But it is rare for negotiations to include demands for system documentation or self-reporting as in technical aviation control. Using input from those to be controlled as a basis for state regulation necessitates the following:

1) Confidence that private parties are loyal to the intention of the system (safety as value).
2) An overlap of interests between regulators and the regulated ("in the same boat").
3) Enlightened self-interest, particularly where confidence in private parties is not very strong.

The likelihood that such conditions are fulfilled is a function of how far-reaching implications of local failures are for the system as a whole. In LTSs the interdependence in the system is high, thus regulation can be based on a different kind of authority relation from the traditional legal-bureaucratic regulation. Indeed, some have argued that the form of control represented by "internal control" can be viewed as a step in a process where state regulation moves away from formal and "top-down" arrangements (Graver 1987).

The traditional *Rectsstaat* is characterized by formal and standardized regulations which private parties experience as material rules. A main concern of such systems has historically been to limit the influence of the state over the private sphere.

While Scandinavian welfare-state regulation is, to a certain degree, governed by general intentions (*fullmakter*) rather than specific legislation, private parties still have to deal with regulation in the form of substantial, often detailed, demands.

"Internal control" may be regarded as a development away from traditional regulation. In addition to regulating through broad intentions, substantial demands are substituted for the procedural demands. In this system, the state gives up being "state" in the traditional sense of the word. Rather it becomes a partner in a process aiming at finding the best solutions. The specific role of the state in such arrangements is to guarantee the framework for rational discourse.

In Norway, "internal control" has also been viewed as the general solution to the problem of regulating complex systems in modern societies. It is a cost-efficient way to regulate, without complex bureaucratic arrangements that do not produce intended results.

REFERENCES

Andersen, S. S. 1988. *British and Norwegian Offshore Industrial Relations*. Aldershot: Gower Publishing.

Carson, W. G. 1982. *The Other Price of Britain's Oil.* New Jersey: Rutgers University Press.

Gustavsen, B., and Hunneius, G. 1981. *New Forms of Work Reform. The Case of Norway.* Oslo: The University Press.

Graver, H. P. 1983. *Sikkerhetsstyring og regulerings teknikk.* Oslo: Sjørettsfondet.

——————. 1987. "Safety Control As A Legal Problem." In *Jahresschrift für Rechtspolitologie*, bd. 1.

Karlsen, J. E. 1982. *Arbeidsforhold på norsk kontinentalsokkel.* Oslo: Oktober Forlag.

Liaaen, O. A. 1982. *Verneombud, tillitsmenn og sikkerhet i oljevirksomheten.* Oslo: Arbeidspsykologisk Institutt.

Olsen, J. P. 1982. *Organized Democracy.* Oslo: The University Press.

FACING THE COLLAPSE OF THE WASHINGTON PUBLIC POWER SUPPLY SYSTEM

Stephen Salsbury
with the assistance of L. Sharon Davidson
Department of Economic History
University of Sydney

This essay is divided into two parts. The first is a factual description of the Washington Public Power Supply System's history. The second is an analysis of why things went wrong. There is no attempt made to find scapegoats: the purpose is to place the Supply System in a larger economic and institutional context. Nor does this work attempt to analyze the very complex legal problems associated with the WPPSS default, this is a major topic in itself.

Historical Background

The Washington Public Power Supply System, or WPPSS (derisively pronounced "whoops!" by the Wall Street financial community), is a major venture by any standards. In 1981 WPPSS had under construction five large nuclear power generating plants each with a rated capacity of about 1,200 megawatts. At the end of that year only fourteen commercial nuclear plants in the United States were operational that had a capacity of more than 1,000 megawatts.[1] The WPPSS annual report of 1981 estimated that $23,787,000,000 was needed to complete its power generating plants. Nearly $5 billion had already been expended.[2] Some idea of the magnitude of the WPPSS project can be gathered by comparing it to the massive federal Columbia River Power System, which alone supplied well over 50 percent of the Pacific Northwest's power. In 1988 this System had a power generating capacity of 12,771 megawatts (compared to slightly more than 6,000 megawatts for WPPSS when completed), which represented an investment of $8,982,371,000. If completed, WPPSS would have had more assets than the largest investor-owned electric utility (The Southern Co—$11.46 billion), the largest railroad (CSX—$7.52 billion), or the largest airline (UAL—$4.04 billion).[3] WPPSS had become a major force in the financial markets. In 1981 the *Wall Street Journal* noted that WPPSS was "the nation's largest issuer of tax exempt bonds."[4]

The Pacific Northwest's Public Power Tradition

WPPSS is a recent organization and its history is intertwined with Pacific Northwest's public power movement. In the New Deal era, the federal government began a massive development of the Columbia River. The aim was to build multipurpose dams for navigation, irrigation, flood control, recreation, and electric power. Two federal entities

T. R. La Porte (ed.), Social Responses to Large Technical Systems, 61–97.

constructed dams on the Columbia and its tributaries: the Bureau of Reclamation and the Army Corps of Engineers. The facilities built by these entities were to be, and still are, operated by them.

In 1937 Congress passed the Bonneville Project Act, creating the Bonneville Power Administration (BPA) which was to transmit and market the power produced at the Bonneville Dam. BPA was required to give preference to publicly owned utilities.[5] The Pacific Northwest already had a strong public power tradition, which was exemplified by Seattle City Light. The emergence of a large federal hydro-generating system encouraged municipal power in the Pacific Northwest and caused the founding of a large number of public utility districts (PUDs) in rural areas. BPA was an intermediary wholesaling power generated by other federal entities to publicly owned (but state-chartered) entities. BPA did not retail power direct to customers, with one exception: it was allowed to sell surplus power direct to large users such as aluminum corporations. BPA could also sell surplus power (that not needed by the preference customers) to investor-owned utilities (IOUs). By 1976 BPA's preference customers consisted of thirty-six municipal systems, twenty-six PUDs, and fifty-three power cooperatives scattered throughout Washington, Oregon, Idaho, Montana, and Wyoming.[6]

For well over two decades the Columbia produced enough hydroelectric power to satisfy the requirements of the region's preference customers and the needs of heavy users, particularly the aluminum industry. In addition, BPA regularly sold power to IOUs within the region and to users outside the Pacific Northwest, always at low rates. BPA had a number of advantages. For one thing, it monopolized one of the best river systems in the United States. For another thing, its power-generating sources were built when construction costs were low and interest rates for government bonds fluctuated between 2.19 percent and 4.21 percent.[7]

Franklin Roosevelt, speaking in 1932, saw federal hydroelectric power projects as a model against which IOUs could be measured. He envisioned "four great government power developments in the United States—the St. Lawrence River in the northeast, Muscle Shoals in the southeast, Boulder Dam project in the southwest, and finally . . . the Columbia River in the northwest. Each one of these," Roosevelt asserted, "will be forever a national yardstick to prevent extortion against the public and to encourage the wider use of . . . electric power."[8] Many public power critics argue that Roosevelt's concept of a yardstick is badly flawed. For one thing, government hydroelectric power projects borrow capital at interest rates below those paid by private enterprise.[9] Furthermore, Columbia River projects were multipurpose and the government deducted from the overall project costs the capital investment devoted to such purposes as navigation, flood control, fish and wildlife, and recreation. Dams built by privately owned utilities also served some of these purposes, particularly recreation and flood control, but IOUs were unable to reduce the cost of their dams as a result of them.[10] The government's allocation of some of the Columbia River System's costs to flood control, navigation, and recreation significantly reduced the cost of the System's hydro plants. BPA's accounts make this clear: as of 1988, the government's total investment in the System exceeded $11,425,000,000. Of this only $8,972,000,000 (or 85.1 percent) was charged to electric power generation: the rest (about 7 percent) was charged to irrigation, which did pay for the capital attributed to it, and the remainder (about 8 percent) was allocated against navigation, flood control, fish and wildlife, and recreation, which produced no revenue.[11] The allocation of these costs lowered the capital charges which BPA's customers had to meet. In addition the government facilities paid no real estate taxes. The results were extremely low rates for all of BPA's customers in the Pacific Northwest whether they were preference utilities, IOUs, or direct service industries.

As the years went on, the importance of BPA and its low rates to the Pacific Northwest became more apparent. In January 1977, a bill for 500 kilowatt hours of residential power, taxes included, was only $7.14 for customers of the Seattle City Light (a BPA preference customer). In nearby Portland, Oregon (a city served by two different IOUs, both of which used substantial thermal generating plants in addition to buying power from BPA) the same amount of electricity cost nearly twice as much, or $13.95. In other cities served by private utilities without access to federal power the bill was higher—$19.46 in Washington D.C., $22.15 in Chicago, $27.29 in Philadelphia, $27.37 in Boston, and a whopping $45.92 in New York City.[12] The direct service industrial customers also benefited. In 1978, BPA served six corporations which operated ten aluminum reduction plants accounting for nearly 30 percent of the United States's domestic aluminum production capacity.[13] In the decade between 1967 and 1977, the aluminum industry alone purchased between a quarter and a third or more of all power sold by BPA.[14] Within the Pacific Northwest federal generating power dominated. In 1976 the Columbia River System accounted for 55.4 percent of all power produced in the Pacific Northwest, as opposed to 22.8 percent generated by IOUs. The rest came from publicly owned operations such as the city-owned Seattle Light. Aluminum alone regularly used 8 percent or more of the entire power output of the Pacific Northwest.[15]

In the 1950s, it was perceived that Pacific Northwest power demand would eventually outrun the Columbia River's capacity. Attention shifted to thermal plants. The situation was complicated by institutional arrangements. BPA had no authority to build any power generation facilities. While the federal government was committed to exploiting the Columbia's remaining hydroelectric potential, there was strong ideological opposition to federal thermal plants. Roosevelt had never argued that power generation and distribution was solely a government function. He stated in 1933 that "as a broad general rule the development of utilities should remain, with certain exceptions, a function for private initiative and private capital."[16] Roosevelt proved a strong friend, however, to public power; in addition to the Columbia River System, an even larger and more powerful organization arose in the South, the Tennessee Valley Authority (TVA). In 1952, however, the Republican party had captured the presidency and the tide of political support for public power was waning. In July 1957, President Eisenhower said in reference to TVA, "By God, if we could do it, before we leave here, I'd like to see us *sell* the whole thing, but I suppose we can't go that far."[17] The question for the Pacific Northwest in the 1950s was, if the federal government would not create more public power, who would? Before this question could be answered outside forces intervened to create the Washington Public Power Supply System.

The Rise of WPPSS

During World War II, the federal government nuclear weapons program established an atomic plant at Hanford, Washington. After the war, the Hanford Atomic Works, controlled by the newly established Atomic Energy Commission (AEC), continued to produce weapons-grade atomic materials. In 1956, Senator Henry Jackson of Washington (a democrat), proposed the construction at Hanford of a reactor to make weapons-grade plutonium 239. He noted that the same reactor could produce steam for electric power generation. The idea was sensible, but the politics were difficult. President Eisenhower and other top republicans feared that this "might be a step toward converting the AEC into a public power-for-sale agency."[18]

In 1958, Congress voted to build the Hanford N-reactor but did not provide any funds for electric power. The entire affair became embroiled in controversy between the IOUs and public power advocates. When public power supporters failed to carry the day in Congress they turned to the State of Washington and found a receptive ear in democratic governor Albert Rosellini. The result was a state municipal agency, the Washington Public Power Supply System, created in 1957. Its owners are public power systems (as opposed to IOUs). At the beginning, seventeen public utility authorities participated in WPPSS. Not all of Washington State's municipal utilities or PUDs were Supply System members; it has never been necessary to be a member of WPPSS to participate in its projects. Nor were participants limited to the state of Washington.

WPPSS is a very specialized entity. It describes itself as "a joint operating agency, not a utility." The organization does not make rates. "Its only purpose is to finance, build and operate power generating plants."[19] From the beginning, WPPSS's backers indicated that it would work closely with BPA. WPPSS's first project, begun in 1962, was the construction of a 27.5 megawatt Packwood Lake hydroelectric scheme, which entered service in June 1964. The facility was financed by a $13.7 million bond issue. The power generated by WPPSS was delivered to BPA which in turn distributed it to the publicly owned utilities which had participated in financing the project. The major breakthrough, which set the pattern for WPPSS's development, occurred on 26 September 1963 when President Kennedy approved a bill authorizing the AEC to sell the N-reactor's by-product steam to the Supply System. Kai Lee and Donna Klemka, in *Electric Power and the Future of the Pacific Northwest*, noted that the bill to provide N-reactor steam for public power resulted from complicated political bargaining. The private utility lobby remained hostile to the AEC entering the field of electric power generation. On the other hand, the Pacific Northwest's IOUs, which had grown used to cooperating with BPA, favored generating power from the N-reactor. The compromise specified that no federal money could be used for the project and that half of the electricity would go to private utilities. This bill set the stage for continued public and private cooperation in developing Pacific Northwest power facilities. It also made WPPSS a key player in regional power development.[20] After the signing of the N-reactor bill, WPPSS arranged a $122-million revenue bond issue to finance a generating plant. The Supply System did not construct the reactor—this was done by the AEC. When the project entered service in 1966, its 790-megawatt capacity made it the largest nuclear power–generating facility in the United

States.[21] Although the power cost was slightly less than 7 mills per kilowatt hour (more than twice the average charge made by BPA to its customers), the amount was still low.[22] The power was distributed by the BPA.

BPA and Pacific Northwest Power Planning

The birth of WPPSS happened while the Pacific Northwest was deliberating over how to meet its growing requirements. The federal government, represented by a number of agencies including the Army Corps of Engineers and BPA, took a major part in these deliberations. Political considerations continued to prevent federal agencies from constructing thermal power generating units. Therefore, if such units were to be built, it would have to be done by state and local public agencies (such as the City of Seattle) and by private IOUs. BPA advised and coordinated power plans but did not implement them. In 1967, BPA convened a joint power planning council with representatives from a large number of utilities. BPA administrator Charles Luce chaired the group. In 1968, the group issued plans for a Hydrothermal Power Program which emphasized cooperation between public and privately owned utilities.[23] The number of plants envisioned under the Hydrothermal Power Program grew rapidly. The first phase included nine projects—three coal-fired and six nuclear.[24] Of the six nuclear plants, WPPSS was responsible for three and Portland General Electric for three (including the Trojan plant). As of 1989 only two of the six nuclear plants—Trojan and WPPSS #2—have been completed and are in operation. The rest have been either cancelled or indefinitely postponed. The second phase of the Hydrothermal Power Program consisted of eleven projects, of which six were nuclear. Two of these were the responsibility of WPPSS and the rest were undertaken by IOUs. All of the second-phase nuclear projects have been terminated, apart from the proposal to continue running the generation plant dependent upon the Hanford N-reactor. Currently the N-reactor has been shut down for safety reasons and there is no firm date for its reactivation.

Those responsible for the Hydrothermal Power Program never thought that most of the nuclear plants would be either postponed or cancelled. Indeed, in the early 1970s there was every indication that all would be required and that they would not come on-line soon enough to meet demand. It is significant that of the five nuclear reactors which formed part of phase 1, four had the active encouragement of the BPA. Only one was entirely the responsibility of private power interests.

The problem for BPA was that its forecasts indicated a Pacific Northwest power shortage by the 1980s and the Agency felt obligated to provide for its preference customers. Unfortunately, the structure of public power in the Pacific Northwest did not lend itself easily to the construction of new facilities. Bonneville could not do it directly, nor were its customers well positioned to create new generating capacity. While a few municipalities, such as Seattle, had their own generating units and could build more, most of BPA's customers were mere distribution systems and relied on BPA for their entire electricity supply. In fact, most BPA customers lacked substantial assets that could be used to underpin the large bond issues needed for nuclear plants. BPA solved this problem through a concept called "net billing."

Under net billing a BPA preference customer could elect to take a percentage ownership in a new power plant. For example, a PUD could buy 10 percent of a 1,200-megawatt nuclear facility. BPA would guarantee purchase of the 120-megawatt power output at a price which would cover the interest payments, debt retirement, and the cost of operations. BPA would give the PUD a credit for this power, then meld it with its own low-cost hydro output. This allowed costs to be spread across all BPA sales, not just those to preference customers. In effect, this required all BPA users to share in the cost of new power facilities—BPA would raise its overall power rates to cover the cost of the new plants. Averaging the hydro with the nuclear power promised the continuation of relatively low-cost power for the preference customers. Another feature of net billing was important: BPA signed "hell or high water" contracts with its suppliers. Consequently, BPA entered into agreements as to the date on which it would start paying for the power. This guaranteed that the interest and debt retirement for the new power plants would be met even if the facilities were never finished. In effect, BPA obligated its customers to take all the risks of a failure or a "dry hole."

In theory the financing was arranged by the public authorities that were "purchasing" generation capacity. This meant that the debt was not Bonneville's but that of the cities, towns, and PUDs. This provided another crucial advantage: state and municipal entities could issue bonds which paid interest that was free of all federal income tax. This tax loophole promised that money could be borrowed at a lower interest rate than could be commanded either by the BPA itself (interest paid on the securities issued by the federal government or its entities is fully taxable under the United States income tax regulations) or by private companies.[25]

Under net billing, it was not even necessary for the power generation project to be initiated by a public authority. The third net-billed project (the WPPSS Packwood Lake hydro plant and the WPPSS Hanford N-reactor were the first two) was the Trojan Nuclear Project sponsored by the investor-owned Portland General Electric Company. In this case, the city of Eugene, Oregon bought 30 percent of Trojan and assigned it to Bonneville under net billing.[26]

BPA's net-billing agreements were made to order for the WPPSS. This organization had little or nothing in the way of assets (only the Packwood Dam and generating plant associated with the Hanford N-reactor) to serve as backing for any bonds it might issue. Nevertheless, WPPSS's success with the Hanford N-reactor and its relationship to public power in the Pacific Northwest made it the ideal vehicle to bring new capacity on line. Consequently, BPA entered into contracts to net-bill power from three WPPSS projects. The first was WPPSS plant #2 at Hanford, which was planned in 1970 and saw ground-breaking in August, 1972. The net-billing agreement for a second project, also at Hanford (WPPSS #1) was signed on 6 February 1973. A third net-billing agreement, WPPSS project #3 at Satsop, Washington, came into effect on 25 September 1973. BPA agreed to net-bill 100 percent of the power for projects #2 and #1 and 70 percent of the power generated by project #3. The remaining 30 percent of project #3 was owned by four IOUs which arranged their own financing: Pacific Power & Light company (10 percent), Portland General Electric Company (10 percent), Puget Sound Power & Light Company (5 percent), and the Washington Water Power Company (5 percent).[27]

Participants in the net-billed WPPSS projects were large in number and varied. As of 1974, there were a total of 109 participants including forty-seven cooperatives, two irrigation districts, twenty-nine municipalities, twenty-six PUDs, and five IOUs. Although the participants were the nominal owners of the WPPSS net-billed plants, they did not issue any securities. Those were issued by WPPSS in its own name but they were in turn guaranteed by the net-billing arrangement. BPA was the key factor enabling the sale of the WPPSS bonds. When the municipal bond market was shaken by New York City's credit problems in 1975–76, BPA ensured that the Supply System could sell its securities at the best possible price. In its 1976 Annual Report, BPA noted that WPPSS bonds had in 1975 been given the highest rating (AAA) by Moody's Investor Services and Standard & Poor's. Unfortunately, WPPSS was unable to sell securities except at "interest rates higher than would normally be expected for bonds with an AAA rating." BPA concluded that this "demonstrated that purchasers did not fully understand how BPA provides security for the bonds under net-billing arrangements." As a result, BPA and WPPSS personnel together attended meetings in over a dozen American cities to emphasize the security provided by BPA through net billing. This allowed WPPSS's bonds to be sold "at interest rates as much as three quarters of one percent lower than pre-1976 WPPSS sales when compared to the standard market indexes for both periods."[28]

In 1973, while the Pacific Northwest's power authorities were planning the second phase of the Hydrothermal Power Program, the Internal Revenue Service prevented BPA from making any more net-billing agreements. Federal taxing authorities ruled that WPPSS securities which were backed by the federal government could not qualify for tax exempt status because they were federal debt in disguise and not true municipals. While the IRS action did not invalidate the previous net-billing agreements, it forced WPPSS to devise new ways to finance additional projects.

WPPSS Decides to Build Units #4 and #5

Through the early 1970s, BPA continued to be pessimistic about its ability to meet the Pacific Northwest's power requirements. As late as 1976, BPA was forecasting regional energy deficits: "For 5 consecutive years—1978–79 through 1982–83—these deficits could exceed 2 million average kilowatts, roughly 2.5 times the electric requirements of the city of Seattle."[29] In June 1976, BPA reinforced its message that the Pacific Northwest faced "continuous energy deficits" by issuing formal notices of insufficiency to all of its preference customers. These stated that for the fiscal year 1983–84 and each operating year thereafter the administration could not meet the preference customers' projected needs.[30]

WPPSS's decision to build two additional nuclear power generation facilities can only be understood in the light of BPA's dire predictions of power shortages. WPPSS approved the construction of units #4 and #5 in July 1974. Although the Supply System had BPA's moral support, it acted without any financial guarantee. WPPSS turned directly to the organizations, public and private, that participated in its units #1, #2, and #3. By July 1976, eighty-eight of the Pacific Northwest's publicly owned utilities had signed agreements to share in the output and costs of the new nuclear projects; a major exception was the City of Seattle.[31] The participants agreed "to pay their respective share of the project's annual cost, including debt service." The agreements stipulated the exact percentage of each project's annual cost and of the output allocated to each participant. Billing was to start when the units entered commercial operation or on 1 July 1988, whichever date was earlier. These contracts also had a hell or high water clause similar to that which BPA had signed in its net-billing agreements with WPPSS. The hell or high water agreement specified that if WPPSS were unable to sell bonds or to pay the principal or interest of the issued revenue bonds when due or was unable to proceed with completion of the units, "each of the participants will pay its proportionate share of the principal due on the revenue bonds together with any other costs associated with the termination of the projects."[32]

Problems at WPPSS

Almost from the start things did not go well for WPPSS. The Pacific Northwest's first nuclear plant built exclusively for the generation of electric power, Portland General Electric's Trojan unit, cost approximately $460 million dollars and came on-line in 1976 "without substantial delays."[33] WPPSS estimated that its first project, #2, would take approximately five years to construct (after ground-breaking) and would come on-line in September 1977.[34] This was not to be. Delays became the order the day. Part of the problem was that the technology used in nuclear power generation was relatively new (this problem is discussed extensively below.) Labor was another trouble spot. Atomic energy as an industry did not exist until the Second World War and there was a shortage of trained technicians and workers. Through 1972, the growth of the nuclear power industry was relatively modest. As of that year, only twenty-six commercial nuclear power plants were in operation.[35] The years 1972 and 1973 found America in the midst of an energy crisis. President Nixon announced on 23 February 1973, that one of the "highest priorities" of his administration during the coming year "will be a concern for energy supplies—a concern underscored this winter by occasional fuel shortages."[36] In October 1973 a war in the Middle East cut off Arab oil supplies. This emphasized the importance of substituting domestic for imported fuel.[37]

The energy crisis gave the nuclear power industry a strong boost. Whereas in 1969 only four new nuclear plants were ordered, the number was forty-one in 1973 with an additional twenty-eight in 1974.[38] In two years—1973 and 1974—America's electric utilities ordered more nuclear plants than were in operation at the end of 1972. The rapid proliferation of new projects ensured labor shortages. Across America utilities hired less skilled, inexperienced workers and this resulted in poor-quality workmanship. WPPSS suffered serious labor problems. For example, in 1975 construction of project #2 slowed when "welds on a pipe whip support bracket tore the inside of the containment vessel."

There followed an extensive investigation, then repairs which were "made on a three shift basis" over two months, and finally over two weeks of testing to ensure the new work met the Nuclear Regulatory Commission's safety criteria.[39] In 1976, work on project #2 encountered major problems with WPPSS's contractors. In January, the contractor responsible for the project's civil construction defaulted and the contractor's performance bond proved inadequate.

The problems of plant design, labor shortages, poor workmanship and strikes caused ever longer delays; five-year construction estimates lengthened to ten- or twelve-year construction periods. Project #2, originally scheduled for completion in 1977, did not come on-line until 1984.

Simultaneously, WPPSS faced rising interest costs. For many years high-grade municipal bonds attracted interest rates under 3 percent.[40] In the 1960s, interest rates began to climb. When WPPSS started building its plants the municipal bond rate was about 5.25 percent. Between 1973 and 1979, the rates for high-grade municipals hovered between 5.18 percent and 6.89 percent. In 1980, interest charges escalated, climbing to more than 11 percent in 1981 and 1982. In those years, the market began to downgrade WPPSS securities and it had to issue bonds bearing interest rates in excess of 14.25 percent! Rising interest rates had ominous implications for a capital-intensive industry such as nuclear power. It was especially serious when the project construction phase lengthened from five to ten or twelve years. The problem was not quite so acute for the net-billed plants since BPA started to pay WPPSS for the power during construction, which enabled WPPSS to meet interest payments as they fell due without borrowing. However, in the case of projects #4 and #5, which were not net-billed, WPPSS had no choice but to capitalize the interest, thus paying interest on interest. This was one reason why the cost of nuclear power plants began to rise so rapidly. Table 1 details the cost estimates for WPPSS's five plants. In 1974, WPPSS estimated that they could be brought on-line for about $4.8 billion dollars. By 1978, the figure rose to $8.1 billion dollars. It was over $15.5 billion in 1980 and nearly $23.8 billion in 1981!

Pacific Northwest Power Projections

If technical problems, labor trouble, and the increase of the cost of capital were not enough, WPPSS also suffered from erroneous power consumption projections. In 1973, WPPSS accepted the judgment of Pacific Northwest power planners, which included BPA, that the region's electricity requirements would grow by 5.2 percent per year.[41] By 1978, the Pacific Northwest's power growth rate had fallen to between 4 and 4.5 percent a year. This decline did not bother WPPSS management, which argued that "even with a slower growth rate, the need for new generating facilities is evident." The revised figures meant, according to the Supply System, that it "could be asked to undertake a project of approximately 1,000 megawatts every two years in the late 1980s and early 1990s."[42] However, the growth rate of power demand continued to decline. In 1980, BPA reported that "for the sixth consecutive year" growth slowed and the rate was now projected to be 3.4 percent per annum.[43] In 1983, BPA issued its second independent

TABLE 1

Capacity, Reactor Type/Builder, and Cost Estimates[a] for Washington Public Power Supply System Nuclear Plants, 1973–82

	WPPSS #2*[b]	WPPSS #1[b]	WPPSS #3[b]	WPPSS #4[c]	WPPSS #5[c]	
Capacity in Megawatts	1,200	1,206	1,240	1,240	1,240	
Reactor Type/ Builder†	BWR/GE	PWR/B&W	PWR/CE	PWR/B&W	PWR/CE	
	Estimated Cost of WPPSS Nuclear Plants *(in thousands of dollars)*					TOTALS (x *$1,000)*
1973	476,000	627,000	749,000	d	d	1,852,000
1974	614,000	990,000	1,000,000	1,009,000	1,210,000	4,823,000
1975[e]	794,000	1,147,000	1,244,000	1,095,000	1,271,000	5,551,000
1976	965,000	1,206,000	1,370,000	1,443,000	1,668,000	6,652,000
1977[f]	1,077,000	1,368,000	1,423,000	1,792,000	1,971,000	7,631,000
1978[g]	1,172,000	1,418,000	1,472,000	1,915,000	2,146,000	8,123,000
1979[h]	1,821,736	2,341,000	2,256,000	2,579,000	2,752,000	11,749,736
1980	2,446,965	2,735,000	3,129,000	3,613,900	3,617,000	15,561,865
1981	3,216,000	4,268,000	4,532,000	5,510,000	6,261,000	23,787,000

Subsequent status

WPPSS #2	May 1984, first electricity generated, commercial operation December 1984; total cost, $3,300,000,000.
WPPSS #1	Project delayed 29 May 1982, 63 percent complete.
WPPSS #3	Project delayed 8 July 1983 for extended period, 76 percent complete.
WPPSS #4	Construction slowed July 1981; construction halted 25 September 1981; project terminated January 1982, 24 percent complete.
WPPSS #5	Construction slowed July 1981; construction halted 25 September 1981; project terminated January 1982, 16 percent complete.

Source: Compiled from *WPPSS Annual Report*, years 1973–82.

*Plants are listed in order of construction; ground-breaking for WPPSS #3 was in August 1972.

†Reactor types are Boiling Water Reactor (BWR) and Pressurized Water Reactor (PWR); reactor builders are Babcock & Wilcox (B&W), Combustion Engineering (CE), and General Electric (GE).

[a]In the case of #3 and #5 these data include costs attributable to participating investor-owned utilities.

[b]Net-billed by Bonneville Power Administration

[c]Not net-billed

[d]Authorized July 1974.

[e]March 1975: major fire at Brown's Ferry TVA Nuclear Plant. December 1975: Trojan Plant of Portland General Electric begins test operation.

[f]25 March 1977: *Science* cites Trojan as one of the nation's most unreliable plants with a capacity factor of less than 20 percent.

[g]March 1978: Trojan shuts down until 22 December 1978. Nuclear Regulatory Commission designates control house walls as substandard.

[h]29 March 1979: Three Mile Island partial meltdown.

twenty-year forecast. The projection varied markedly from those produced when WPPSS embarked upon its massive program to build nuclear power faculties. The BPA forecast made three different assumptions: a low growth estimate, a base estimate, and a high estimate. Even the high estimate only projected a power demand growth of 2.6 percent per year, while the base was 1.8 percent and the low forecast 1.3 percent. If the low forecast estimates were correct, the Pacific Northwest could look forward to an energy surplus into the twenty-first century. Even the base forecast saw surpluses existing until 1989. Only the high forecast saw substantial energy deficits but these would not begin until after 1986.[44] An indication of how far off the power estimates for the Pacific Northwest were can be gleaned from the following facts. As of 1989, of the twelve nuclear plants projected in phase 1 and phase 2 of the Pacific Northwest's hydrothermal program,[45] only two—Trojan and WPPSS project #2—had come on-line; all the rest had been cancelled or delayed. In contrast, most of the coal-fired projects were completed but some had extreme difficulty finding markets for their power. A case in point is Colstrip #4 which came on-line in April 1986. Montana Power, which owned 30 percent of the 700-megawatt unit, has found it difficult to dispose of the output at any price. In 1988, the company reported that it was selling power from its share of Colstrip at prices that did not cover all of the costs. It did manage a six-year sale of 20 megawatts of capacity to Pacific Corp. (the old Pacific Power & Light) commencing in 1990. Finally, on 17 July 1989, Montana Power was able to sell 160 megawatts of Colstrip power under a 21.5-year contract to the City of Los Angeles's Department of Water and Power. Furthermore, the company still has 74 megawatts of power for which it cannot find a market.[46]

WPPSS Runs Out of Money

The year 1981 found WPPSS in turmoil. The major problem was the financing of units #4 and #5, which were not net-billed. Events outside WPPSS's control had cast doubt over the units' future. The Federal Reserve Bank, to dampen inflation, instituted high interest rates which pushed the United States, and particularly the Pacific Northwest, into a sharp recession. The suppliers of basic raw materials, such as aluminum, suffered especially. Pacific Northwest power demand tumbled, casting doubt on previous power consumption forecasts. Furthermore, the combination of the energy shortage following the Arab oil embargo of 1973 and the ecology movement forced a general reevaluation of energy policy. Cheap energy, hitherto considered a virtue, became identified with waste. In the Pacific Northwest, a powerful, newly emerging conservation movement charged that energy production created substantial environmental damage.[47] The 1970s also witnessed a public reaction against nuclear power, particularly after the serious fire at TVA's Brown's Ferry in March 1975, and the partial melt-down at Three Mile Island in March 1979.

The growing influence of the conservation movement in the Pacific Northwest was reflected in the Pacific Northwest Electric Power Planning and Conservation Act passed by Congress and signed into law on 5 December 1980.[48] This act, while it provided a mechanism for BPA to acquire power from non-federal resources such as WPPSS, specified that before new power was purchased BPA's administrator must first attempt to decrease power use. The law's supporters stressed that conservation had been "ignored" and that "the opportunity for conservation of electric power in the [Pacific Northwest] region is great."[49] In anticipation of the enactment of the law, BPA had already embarked on a number of important power-saving initiatives including a home weatherization program, solar hot water heating, and the introduction of more efficient irrigation pumps.[50]

The Pacific Northwest's declining power requirements especially endangered WPPSS units #4 and #5. Whereas BPA agreed to purchase the entire output of WPPSS first three nuclear reactors, there was no such guarantee for #4 and #5. The only certain market for these reactors was that of the participating utilities. The rapidly escalating project costs ensured that such power would come on-line well above traditional Pacific Northwest rates. In a situation where the Pacific Northwest had surplus power, the WPPSS participants might find themselves forced to take high-cost power under the hell or high water contracts they had signed. The investment community began to worry about what might happen if the participating municipalities and PUDs were forced to take high-cost power (or worse yet, pay large charges for no power) under the hell or high water clauses. In an article in June 1981, the *Wall Street Journal* observed that there had never "been a significant [court] test of a hell or high water bond."[51] These sentiments made it more difficult to sell bonds for units #4 and #5.

There was another obstacle. WPPSS had become the largest municipal bond seller in the United States. Most trust accounts, retirement funds, and mutual funds had limitations on the amount of securities that they could or would buy from any one issuing authority. By 1981, Wall Street was awash with WPPSS paper and the market was fast drying up for its securities. In March 1981, WPPSS went to Wall Street selling $200 million of bonds to continue the construction of units #4 and #5. These securities were rated A-1 by Moody's and A+ by Standard and Poor's, which were lower than the AAA rating given the BPA-backed net-billed projects #1, #2, and #3. This turned out to be the final sale of securities for #4 and #5. In March, the financial world still assumed that the total cost of all of WPPSS's five nuclear plants would be about $15.9 billion. Within three months, WPPSS's Managing Director Robert Fergerson announced that his new estimate put the total cost at $23.9 billion. The cost of #4 and #5 alone had escalated to more than $11.7 billion.[52]

By July 1981, WPPSS was at a crisis point. The system was fast running out of money for #4 and #5 and either had to sell new bonds or stop work. In July Merrill Lynch issued a research report which warned that WPPSS might not be able to raise any more capital in the bond market and that the status of outstanding bonds was "in serious jeopardy."[53] WPPSS's immediate reaction was to sell nuclear fuel purchased for #4 and #5 to its net-billed projects. This gave WPPSS about $100 million which allowed construction to continue for another sixty to ninety days. Some Wall Street authorities suggested that WPPSS might use this grace period to convince BPA to extend to #4 and #5 the same guarantee that it made for #1, #2, and #3. Merrill Lynch argued against this, suggesting that such a strategy would most likely "imperil the strength of securities behind the bonds of the other three plants through a dilution of BPA's financial operations."[54] In any case, BPA would have found it difficult to make an immediate guarantee to purchase power from #4 and #5 because of the conservation provisions of the Pacific Northwest Electric Power Planning and Conservation Act of 1980.

In the midst of the crisis, the governors of Washington and Oregon named a blue-ribbon panel to advise on what should be done. The panel recommended a halt in construction on #4 and #5 and the "mothballing" of work that had been done in the hopes that it could be restarted at a later date. At about the same time, the Washington State Legislature revised the WPPSS Act to bring into its management business leaders with broad experience. The new arrangement created a revised executive board for WPPSS consisting of eleven members—five nominated by the WPPSS board, and six outsiders, three of whom were appointed by Washington's governor and three by the WPPSS board. The administrative change did not provide any money. WPPSS estimated that preservation costs would run between $150 million and $180 million. Many had no confidence in these figures since WPPSS's record in forecasting costs had been so poor. In addition, WPPSS's participants were liable for interest payments on the $2.25 billion in bonds already issued to finance the two units. Raising additional capital to finance preservation of the partly constructed projects proved impossible. IOUs such as Pacific Power & Light feared that state utility regulators would prevent any additional money invested in the abandoned nuclear plants from being added to their rate bases. Many of the PUDs feared that not all would participate in preservation costs, thereby throwing greater burdens on those that did. In January 1982, WPPSS's outside directors Charles Luce (chairman of Consolidated Edison), Edward Carlson of UAL, and Michael Berry (former president of Seattle's Seafirst Bank), resigned out of frustration and fear that they might be held liable for WPPSS's problems.[56] The participants could not agree on a plan to preserve units #4 and #5 and on 22 January 1982, WPPSS formally abandoned the projects.

WPPSS Bond Default

At this point WPPSS's structure worked against it. Private companies have substantial revenue-producing assets and are allowed by state regulatory commissions to earn a profit on their investments. Normally, they pay a large percentage of their profit in dividends to investors. However, if things go wrong private utilities can cut dividends and divert the cash flow to meet the interest payments on money borrowed to finance a failed or delayed investment. A fully vertically integrated public authority, such as the TVA, also has large revenue-producing assets and the ability to raise power rates in order to generate cash flow to help them to pay the interest on money borrowed to finance an unproductive investment. For example, in 1982 the Cincinnati Gas & Electric Co. (CG&E) was in charge of building the Zimmer Plant, which used a General Electric boiling water reactor similar to that installed in WPPSS unit #2. Zimmer had many of the problems that WPPSS experienced—faulty construction, long delays, high interest rates, and rising costs. In 1984, CG&E and its participants scrapped the nuclear part of Zimmer and converted the station to coal. The state regulatory agencies forced Zimmer participants to write off $861 million invested in the abandoned nuclear reactor against their profits. CG&E, because of its strong earning base, accomplished this without the disruption of dividends or default on debt.[57] WPPSS had no cash flow which it could use to meet the interest payments and amortize $2.25 billion of debt on units #4 and #5. Its only revenue-earning assets were the small Packwood Lake project and the generating plant associated with the Hanford N-reactor. Both of these were financed through net-billing arrangements with BPA which precluded raising additional revenue for other purposes. WPPSS could only look to its participating utilities to service and amortize the debt on units #4 and #5. WPPSS's problems caused great concern in the Pacific Northwest. Power authorities had told citizens that participation in WPPSS projects ensured low-cost power. Residents were horrified to learn that the reverse might be true. For example, the town of McCleary, Washington, through its publicly owned McCleary Light & Power, had a 0.00123 percent stake in plants #4 and #5. The town's liability for the principle alone was almost $2.8 million or approximately $3,700 per resident.[58] McCleary's residents faced the prospect of massive rate boosts to pay off the bonds for plants that would never operate. This was in addition to large rate increases instituted by BPA because of its net billing for the WPPSS projects #1, #2, and #3. In 1979, BPA had instituted rates which increased its revenue by 88 percent. In 1981, BPA raised its rates for preference customers by 59 percent. A further 60 percent increase for preference customers occurred in 1982 and in 1983 BPA proposed another 45 percent increase, which was finally cut to 22 percent.[59]

The WPPSS crisis triggered a strong reaction in Washington State. A group called Don't Bankrupt Washington sponsored a referendum containing an initiative that would require voter approval of any bonds issued by WPPSS after 1 July 1982. Contractors and investment bankers who had sponsored the previous bond issues spent nearly $1.5 million to defeat the initiative. Nevertheless, on 3 November 1981 the initiative passed handily.[60] While a court eventually decided that the initiative could not apply to WPPSS projects under construction prior to the passage of the law, the vote in favor of the measure indicated public dissatisfaction with WPPSS.[61] Ratepayer unrest in the Pacific Northwest posed a potent threat to the WPPSS "take or pay" bonds issued to finance the construction of projects #4 and #5. To clear the air the Chemical Bank, which acted as trustee for the WPPSS bondholders, filed suit in the King County, Washington Superior Court seeking a declaratory judgment that would validate the contracts. Although WPPSS was nominally the defendant, the suit aimed to bind the municipal and public utility districts to pay the interest and amortization costs of $2.25 billion bonds.[62] During the suit the participants made no payments but awaited the judicial decision as to the validity of the hell or high water agreements.

Finally, on 15 June 1983 Washington's State Supreme Court voided the agreements, determining that the municipal utilities did not have statutory authority to enter into them. On 6 November 1984, the State Supreme Court reaffirmed this ruling and invalidated all of the remaining participants' agreements including those entered into by the IOUs. The Chemical Bank appealed the Washington State Supreme Court's rulings, but on 29 April 1985 the United States Supreme Court declined to hear the case.[63] The court rulings forced WPPSS to default on the bonds for projects #4 and #5. The default took place on 22 July 1983, and in August the Chemical Bank, as trustee for the bondholders, moved to take charge of the business and properties of nuclear projects #4 and #5.

The Curtailment of the WPPSS Projects

The default on units #4 and #5 destroyed WPPSS's ability to sell more bonds. This was true even for the net-billed projects whose securities were guaranteed by BPA. Overriding this agreement was the Wall Street reaction to WPPSS's predicament. In May 1983, *Business Week* suggested that the default on units #4 and #5 could force WPPSS itself into bankruptcy, threatening the $6 billion of bonds issued for the other three plants. The journal even postulated that BPA, because of financial woes caused by the downturn in power demand, might be unable to meet its net-billing obligations.[64] *Fortune* pointed out another problem: the bond-holders of the defaulted units might try to "grab all WHOOPS assets" including plants #1, #2, and #3.[65] Because WPPSS had twinned unit #1 with unit #4 and unit #3 with unit #5 (twinned units had considerable property in common), the creditors of units #4 and #5 claimed part of the assets of the other projects. The market took this threat seriously. Bonds for project #3, issued in February 1981, and bearing 11.125 percent interest sold in June 1983 at less than 70 percent of face value.[66] Inability to sell bonds stopped construction on units #1 and #3. Work continued only on the nearly completed unit #2, which did not need a new bond issue and was financed by the cash flow from the net billing. Legal problems tarnished WPPSS's other securities for many years and the annual reports contained notes warning that the creditors of projects #4 and #5 were taking action or could take action to obtain assets from other WPPSS projects.[67]

Not only did the default halt construction on units #1 and #3, but it also prevented the refinancing of high interest debt issued for units #1, #2, and #3. Both of the major Wall Street credit rating agencies, Moody's and Standard and Poor's, abolished the credit rating on all WPPSS's bonds in May 1983. Only in November 1988 did Standard and Poor's reestablish the ratings for the securities issued for units #1, #2, and #3. This was possible because Congress amended the municipal bankruptcy laws strengthening "the legal wall between WPPSS projects" and minimizing the possibility that the creditors of units #4 and #5 could attach the assets of #1, #2, and #3. While the poor market for electricity in the Pacific Northwest continues to prevent construction from restarting on units #1 and #3, Standard and Poor's AA- rating for the bonds of units #1, #2, and #3 does make it possible for the system to consider refinancing $2,254,925,000 of high-interest bonds in the future.[68] However, at this writing the start of construction on mothballed plants #1 and #3 seems a long way in the future.

Analysis

Institutional Arrangements in the Pacific Northwest

The major causal element in WPPSS's failure was its badly flawed institutional structure. Electric power systems have always been both capital-intensive and technologically complex. Thomas P. Hughes, in his book *Networks of Power: Electrification in Western Society, 1880–1930*, notes that electric power systems "consist of power generation, transformation, control, and utilization components [such as street railways] and power transmission and distribution networks."[69] In the United States, much of the electrical utility industry emerged as systems containing most or all of the relevant functions. For example, in the WPPSS area the Washington Water Power Company, an IOU, not only generated its own power but transformed this power for various purposes including industrial and home use. Washington Water Power developed a large distribution network, and for many years owned and operated a streetcar system.[70] The vertically integrated structure of the Washington Water Power Company became the norm for most American private utilities.

As the utility industry grew a need arose to transmit power between the various independent IOUs and to coordinate this activity. One of the first large planned coordinating systems started in the 1920s. The genesis was the Susquehanna River's hydroelectric potential at Conowingo, Maryland. At the time it was built, the Conowingo Dam produced more electricity than any other American hydroelectric facility (with the exception of Niagara Falls) and generated more power than could be effectively used by a single utility. To solve this problem the dam's builder, Philadelphia Electric, took the lead in the creation of a Pennsylvania/New Jersey power pool which linked the Conowingo hydro project with three IOUs—Philadelphia Electric, Pennsylvania Power & Light, and the Public Service Electric & Gas Company of New Jersey. The resulting power pool played an important role in maximizing the efficiency of the Conowingo project, particularly because power loads peaked at different times in the three systems thus enabling the shift of power from one to the other as needs arose. This pool became a model for similar arrangements elsewhere in the United States. Significantly, the pool was not a separate entity but the conjunction of three powerful, and well-financed, vertically integrated IOUs.[71]

The power system in the Pacific Northwest developed very differently from that in Eastern Pennsylvania, Maryland, and New Jersey. The Pacific Northwest's power structure was largely unplanned and developed out of the public power controversy during the New Deal. Roosevelt's main public power interest was not the Pacific Northwest, but the Tennessee Valley. Under Roosevelt's encouragement the federal government created the Tennessee Valley Authority to develop the electric power potential of the Tennessee River. Unfortunately, private companies already served the Tennessee Valley and there was a power surplus. The TVA immediately came into conflict with the major IOUs, most of which were controlled by a holding company, the Commonwealth & Southern Corporation, presided over by Wendell L. Willkie.[72] In 1939, after much controversy, Commonwealth & Southern Corporation sold its Tennessee Electric Power Company to the TVA.[73] The power fight in the Tennessee Valley created a government authority that resembled an investor-owned utility. The TVA built and owned large-scale generation units, constructed substantial transmission facilities, and sold power through its own retail outlets and to other utilities both public and private.

Pacific Northwest power development took a very different path. Whereas IOUs in the Tennessee Valley had fully saturated the market for electricity and had to be pushed aside, the Pacific Northwest already had a strong municipal utility tradition in the cities of Seattle and Tacoma. These provided a ready market for federal power. Furthermore, rural power was underdeveloped and the Northwest was a good place to establish county-based PUDs. Public and private power found it easier to coexist in the Northwest and the BPA had little resemblance to the TVA. The Pacific Northwest saw no confrontation of the kind that TVA had with the Commonwealth & Southern Corporation. There was no Columbia River Authority. In a spirit of compromise with private power interests, New Deal public power advocates used the Army Corps of Engineers and the Bureau of Reclamation to build generating facilities. BPA did not operate the plants, nor did it distribute electricity to retail customers. It was merely the middleman through which federal hydropower was transmitted to going municipal utilities and newly emerging PUDs and cooperatives. Unlike the Pennsylvania/New Jersey power pool, which was closely connected to its parent investor-owned utilities, BPA had weak links to the other parts of the Pacific Northwest power system, the municipal utilities, the PUDs, the IOUs and even the Bureau of Reclamation and the Army Corps of Engineers. BPA never developed the variety and depth of management skills which characterized the TVA.

Timing was also important. Federal Columbia River power first came on-line in 1937. After the Second World War began in 1941, BPA sold an increasing amount of surplus federal energy to direct service customers, particularly the aluminum smelters.[74] In the postwar era, many public power advocates hoped that the fragmented power system which came into existence with BPA in 1937 would be converted into a TVA type of operation. But the private power interests were too strong. The election of a Republican administration in 1953 which was hostile to federal power ensured that BPA would be unable to play an effective role in planning, financing and developing new generating capacity in the crucial decades of the 1950s and 1960s.[75]

WPPSS was born out of BPA's inability to assume the role that TVA had undertaken in the Tennessee Valley. WPPSS's defects were the fault of neither the BPA nor its administrators, but resulted from the complex economic and political legacy of the New Deal. WPPSS is a very specialized organization. When the State of Washington created it in 1957, it made WPPSS a political subdivision of the state. BPA had no direct control over it. WPPSS was in theory responsible to the state's municipal electric systems and PUDs. WPPSS compared itself to "a city with a council-manager form of government. The city council sets policy and authorizes projects; the city manager carries them out with the city staff."[76] WPPSS management was vested in a board of directors consisting of one representative each from a number of PUDs and municipal electric systems in Washington State.[77] For example, in 1973 WPPSS directors consisted of twenty-one individuals of whom eighteen came from PUDs and three from municipal lighting systems. With the exception of Seattle Light and the utilities division of the City of Tacoma, all of these districts were small and few were vertically integrated. Most merely distributed power received from BPA. Only WPPSS representatives from Seattle and Tacoma had management experience in running large organizations or in facing the complex decisions required in a fully integrated electric power system.

Not only did the WPPSS board lack experience but it also lacked the kind of staff which would normally be associated with a complex technical organization. In 1970, when WPPSS was drawing up plans to launch its first nuclear venture, project #2, WPPSS only had sixty-six employees. The staff only totalled ninety-one at the beginning of 1973 when WPPSS was approving the building of units #1 and #3. Of this number, forty-two were assigned to the operation and maintenance of the Hanford Steam Plant and the Packwood Lake hydroelectric project, which meant that at a time when vast plans for nuclear expansion were under consideration WPPSS had a staff of less than fifty at its headquarters, some of whom were secretaries and other support people. It was not until the end of 1973 that WPPSS central staff expanded to 222. WPPSS hired most of its staff with nuclear experience simultaneously with the decision to build two additional nuclear plants.[78] At first glance WPPSS might be criticized for undertaking large complex technical projects with an inexperienced staff. Yet the WPPSS institutional framework virtually required that it follow the path it did. WPPSS had no revenue whatsoever in 1970, except the proceeds of the net billing of the Packwood Lake and Hanford N-reactor, which were only enough to pay minor administrative costs, and the interest and amortization of the debt of those two projects. WPPSS had no taxing power. It therefore had to engage in extensive advance planning for its nuclear plants with a small, inexperienced staff. Only after plans were made and approved could bonds be sold which would provide enough money to acquire suitable staff with nuclear experience.

Technical Constraints

WPPSS's tiny and comparatively inexperienced staff was a severe handicap. While the production of electricity by ordinary means—that is by water power, or by fossil fuels—was well understood in the 1960s, the opposite was true of nuclear stations. Prior to 1970, the year in which WPPSS planned its first unit, there were only twelve commercial nuclear power plants in the United States. Most of these were small—four of the plants had a megawatt rating of less than 70. Only three of the plants had a megawatt capacity of more than 600 and this number included the Hanford N-reactor.[79] Starting in the late 1960s, electric utilities began to order a new series of large nuclear reactors most of which had capacity ratings of 1,000 or more megawatts. These units were of new design and had not been tested in commercial operation. The WPPSS units fell into this class.

Gregory Allen Bischak has analyzed the important topic of safety in nuclear power generation, and has identified three stages of risk assessment. The first, *ex ante* stage was "based on safety research conducted prior to the commercial deployment of the technology and involved basic research and prototype testing conducted from 1947 to 1960." The second—"concurrent risk assessment"—took place "from 1961 to 1974 and was based on the practice of researching and testing new safety systems while simultaneously deploying these safety systems on scaled up commercial reactor designs." The third category, *ex post* risk assessment, "occurred from the mid-1970s onward and was based on negative feedback from actual operating experiences and operational R&D aimed at improving safety and performance."[80]

WPPSS started its nuclear expansion during Bischak's second phase, the most risky possible time. The federal government belatedly recognized the dangers of nuclear power when, in 1974, it created the Nuclear Regulatory Commission (NRC) to oversee the industry. In March 1975, a serious fire at the TVA's Brown's Ferry, Alabama nuclear reactor caused deep concern about the safety of large-scale nuclear plants.[81] The incident happened at a General Electric boiling water reactor similar to that being installed by WPPSS at its plant #2. The fire broke out when an electrician "held a small lighted candle near some plastic insulation to find out whether air was leaking into an area under the control room." Although the fire caused no release of radioactive material, it destroyed 1,600 control cables, some of which were used to operate the plant's safety devices. The investigations focused on the plant's design; it was alleged that the Atomic Energy Commission had approved the Brown's Ferry plant "even though it apparently did not meet regulation requirements." Investigators criticized the flammable material used in the Brown's Ferry units.[82] The Brown's Ferry incident caused the NRC to review all plans for nuclear plants and require changes in design to eliminate the flaws discovered.

GE's boiling water reactor had other defects. In the early 1970s both the CG&E Company and WPPSS made large expenditures to redesign the system for containing radioactive materials. In the case of CG&E the cost exceeded $360 million. Both CG&E and WPPSS filed law suits alleging that GE had mislead them about the safety of its nuclear reactor design.[83]

The redesign of nuclear plants already under construction caused by the Brown's Ferry incident and the problems with steam generation in the GE boiling water reactors were only two of a series of engineering challenges facing WPPSS. These cases emphasized the technical complexity of building nuclear reactors and the need for a highly trained staff. In a 1977 *Science* article on nuclear power, Deborah Shapley described the Connecticut Yankee reactor, which came on-line in 1968, as a success. This reactor, a pressurized water model, had been built "on the basis of a technology which received extensive operation testing." It was engineered by the Yankee Atomic Electric Company, an organization formed in the 1950s under the auspices of a group of New England utilities solely to "design, build and operate" a demonstration project, "Yankee Rowe."[84] The contrast between the Yankee group and WPPSS could not be more striking. In the Yankee case several well-financed IOUs developed an experimental 175-megawatt nuclear power station and then used their experience to build the larger Yankee plant. Glenn Reed, who had been associated with the Yankee atomic program, commented that where utilities did not have highly trained experts to supervise nuclear plant design and construction, the result was usually a poorly performing power plant.[85] Shapley noted that nuclear plants of the Sacramento Municipal Utility District and the Portland General Electric (neither one of which had prior experience with nuclear power) were among the worst performing in the nation.

WPPSS's institutional arrangement gave it another serious disadvantage. Under state law WPPSS had to put all contracts out to tender and accept the lowest bid. This rule made sense for normal city responsibilities such as garbage collection. It was also appropriate for purchasing standard commodities such as water pipes, copper wire, and automobiles. But price was not an adequate guideline when buying nuclear reactors. Complex technology had plagued the electrical industry for many decades. Harold Passer in his book on the electrical manufacturers noted that in the 1890s technologically ignorant street railway executives purchasing electric motors often insisted on tests that told little about the quality or appropriateness of electric motor design.[86] Washington State's bidding laws were not of negligible consequence: they were at the root of WPPSS's decision to buy three different kinds of nuclear reactors for its first three plants! Unit #2, the first to be ordered, used a General Electric boiling water design. Plant #1 had a Babcock and Wilcox pressurized water reactor, and plant #3 contained a Combustion Engineering pressurized water reactor.[87] It seems inconceivable that a small municipal organization, with little capital and a small staff, could undertake simultaneously the construction of three large nuclear reactors of radically different design. In contrast, the TVA built its first three units at a single site, Brown's Ferry. TVA ordered identical GE boiling water reactors, which allowed the utility to capitalize on the experience of building the first unit in the construction of the other units. The TVA's financial strength and the depth of its management was undoubtedly one of the reasons why it successfully completed all three reactors. (The first, which began construction in 1967, achieved full power in 1974. The last, which started construction in 1968, came on-line in 1977.) Despite the Brown's Ferry fire of 1975, TVA erected the units for less than $900 million.[88] This was less than one-third of the total cost of WPPSS's only operating unit, #2, which had a total cost of $3.3 billion.[89]

WPPSS Managerial Decisions

While it is true WPPSS's management lacked depth and financial resources, could it have avoided the collapse of its program? The answer involves a study of the management decisions and the data which WPPSS had at its command. The sources do not include manuscript material. Rather the analysis rests on WPPSS's annual reports, those of BPA, and information available in the financial press. The annual reports reveal what management claimed it knew, and the financial and technical press of the day reveal what any alert, intelligent management ought to have known. The lack of access to WPPSS's internal documents probably favors management since this material may reveal knowledge of problems that were covered up or ignored.

It must be emphasized that the WPPSS projects did not fail because of opposition to nuclear power or environmental concerns. WPPSS collapsed primarily because it ran out of money. This was because WPPSS proved unable to contain costs. WPPSS made three sets of decisions. The first decision—to build unit #2—was made in 1970. The second—to construct units #1 and #3—took place in the first part of 1973. The third decision—to build units #4 and #5—was made in July 1974.[90] Despite its peculiar institutional structure, WPPSS had a very large capacity to raise capital. It sold more than $9.3 billion of bonds of which over $7 billion was for the net-billed projects.[91] Under certain circumstances WPPSS might have sold even more bonds. It will be recalled that it was the non–net-billed projects which brought WPPSS down. The legal complications surrounding WPPSS's default on these bonds undermined all of its securities. Had all projects been net-billed it seems likely that WPPSS could have terminated #4 and #5 without risking units #1 and #3.

In 1970, when WPPSS made its initial decision to build a nuclear plant, nuclear power seemed relatively inexpensive. The early, small nuclear units had come on-line without incident and were giving good service. Construction time for nuclear facilities was about five years and the cost per unit was well below $500 million. Portland General Electric's Trojan unit started up in late 1975 at a cost of $460 million. It is unreasonable to assume that in 1970, or even in 1972, when ground was broken for unit #2, WPPSS's management could have recognized that a new era had dawned. Only hindsight can determine that the period of "concurrent risk assessment," which drastically revised nuclear safety standards and caused massive alteration in the plans for new nuclear plants, began in 1961. WPPSS's $476 million projected cost for its first unit seemed reasonable. Had WPPSS limited itself to this project it would have been criticized for cost overruns but there would have been no bond defaults. In fact WPPSS did complete #2 in 1984 at a cost of $3.3 billion.

The decision to build units #1 and #3 is more open to question. By then construction had started on #2 and cost estimates for nuclear projects had begun to escalate. WPPSS costed the new projects at $627 million and $749 million respectively. These amounts were well within WPPSS's borrowing power, but management should have asked why, in a single year, there had been such a large increase in the price of nuclear plants. However, strong forces urged WPPSS forward. BPA insisted that the Pacific Northwest was facing an acute power shortage. Furthermore interest rates for municipal bonds, which had been about 5 percent since 1969, seemed reasonable. BPA's extension of net billing to units #1 and #3 reinforced WPPSS's decision.

WPPSS's major error occurred in July 1974 when it proceeded with another two units. By this time warning signs were clearly visible. First, there was the matter of money. Unit #2's estimated costs had reached $614 million and WPPSS was estimating about $1 billion each for units #1 and #3. The projections were greater for the new units—#5 was costed at over $1.2 billion. Second, WPPSS was no longer innocent of the problems of nuclear plants. Its 1974 report told of troubles with unit #2: "As originally scheduled, . . . construction was to have been nearly 39 percent completed by the end of 1974" (when in fact the project was only 13 percent complete. The major causes of delay were "skilled manpower shortages, problems in obtaining equipment, quality control programs . . . , numerous design changes, [and] planning problems experienced by contractors."[92] J. J. Stein, WPPSS's managing director, indicated that an important cause of the delay was "the need to make design modifications to comply with changing local, state, and federal requirements."[93] By 1974, WPPSS's management had already fingered what became an increasingly severe problem for all nuclear plant construction in the 1970s—that WPPSS was putting plants out to contract before the final design was complete. Additionally, government agencies were bringing in new regulations which required alterations of supposedly firm design plans. This allowed contractors to make low bids knowing that agreements could be renegotiated in midstream when design specifications changed. All of this was well known to WPPSS even before the fire at TVA's Brown's Ferry units. There were other warning flags as well. BPA could not offer to net bill the power for units #4 and #5, making the financial risk much greater. In addition, interest rates were rising.

These facts should have indicated caution but they seem to have had the opposite effect on WPPSS's directors. WPPSS apparently believed that because the cost of nuclear power was rising it should enter into contracts to lock in current prices. Consequently, before participants could be signed up for projects #4 and #5, or financing could be arranged, WPPSS committed itself to spending more than $230 million on equipment and nuclear fuel for the units. The use of contracts to contain inflation made sense if purchasing standardized products, but it was not valid when the contracts were for plants for which the specifications were continually evolving. WPPSS's strategy, instead of containing costs, paved the way for endless litigation. For example, in 1976 WPPSS sued the contractor in charge of unit #2's civil construction for breach of contract. The contractor in turn filed suit against WPPSS asking for damages of $24.5 million. In turn, a subcontractor of the original contractor filed suit against the System for $11.9 million and punitive damages of $20 million. This was typical of the problems which escalated throughout the 1970s.[94]

The years 1975 and 1976 were in retrospect the most critical for WPPSS. This is because it might have been possible to terminate construction on projects #4 and #5 since relatively little had been spent on them. As of 30 June 1975, only $8 million had been spent on the two projects; this had increased to about $38 million by 30 June 1976.[95] While a further $372 million had been committed for the two projects, most of this was for nuclear fuel and hardware that could have been used at projects #1 and #3. In fact, as previously noted, when WPPSS reached a financial crisis in 1981 it managed to sell $100 million of nuclear fuel destined for projects #4 and #5 to the net-billed units. In hindsight, the eighteen months between 31 December 1974 and 30 June 1976 should have provided management with a foretaste of what was to come. Cost estimates for unit #2 rose from $614 million to $965 million. Percentage rises on the other units were not so great but the overall estimated costs of the five units rose from $4.82 billion to $6.65 billion.[96] The Brown's Ferry fire in March 1975, which received substantial publicity in the nation's financial press, demonstrated that nuclear reactors, especially those of the type that WPPSS was installing in its unit #2, had substantial design flaws. The *New York Times* on 26 March 1975 reported that the NRC had determined that the Brown's Ferry fire made several of the "safety systems inoperative." The NRC asked utilities to “review the design of walls and floors of control rooms with particular attention to flammability and to examine procedures for controlling construction work where it might affect operating reactors.”[97] WPPSS Managing Director Stein stated in his 1975 annual report that "the need to make substantial revisions to plant design to meet new guide-lines and codes adopted by the U.S. Nuclear Regulatory Commission" had delayed construction.[98] It was also in 1975 that WPPSS discovered that the faulty design of the GE boiling water reactor's containment vessel might allow “accidental releases of large volumes of steam.” It was therefore “necessary to design strengthening for the vessel.” Furthermore the NRC “required a design change in the sacrificial shield wall that will be installed above the reactor vessel as additional protection against impact.”[99]

Labor problems were also evident in 1975 and 1976. A prolonged strike shut down construction for five months. There were also serious problems with the quality of labor. When WPPSS construction commenced in the early 1970s, experienced welders were in short supply. Local 598 of the United Association of Journeymen and Apprentices of Plumbing and Pipe Fitting, which served the WPPSS region, had fewer than 500 members. Almost all were enjoying steady work. Eventually, the Local recruited ten times this number, many of them welders, but only by attracting nomadic welders who travelled from site to site. In the 1980s, the quality of welding became a serious issue not only for WPPSS but at a number of other plants including CG&E's Zimmer plant.[100] This problem haunted WPPSS after the partial melt-down at Three Mile Island in March 1979. After that incident the NRC doubled its surveillance of reactors under construction. An examination of WPPSS unit #2 disclosed violations of NRC rules for which the Commission assessed a $61,000 civil penalty. The trouble was “faulty welding and concrete work on the sacrificial shield . . . enclosing the reactor.” This work had been done several years previously, probably in 1976.[101]

Despite the many warning signs, in 1975 and 1976 WPPSS directors seemed never to consider a delay of construction on units #4 and #5. Institutional arrangements worked against any such strategy. Bonds were sold under agreements that required separate accounts for each project. This created a wall between the projects that made it difficult, if not impossible, to shift funds from one to the other. Furthermore, as previously noted, WPPSS net-billing agreements did not generate a cash flow that could be used for purposes other than providing for the operating costs, debt service, and bond amortization of the specific unit. The termination of the net-billed projects #4 and #5 would have caused particular problems since the cost would have been passed on to the participating utilities. WPPSS officials may have felt that forcing participants to raise rates for delayed projects might undermine faith in the entire program.[102] Although curtailment of expenditure on units #4 and #5 was essential to WPPSS survival, management refused to consider it. In 1975, Ed Fischer, chairman of WPPSS executive committee, argued that the agency's projects were essential in helping America solve its energy crisis. Manager J. J. Stein recognized the problems caused by changing safety standards. His solution was “a slow-down in the implementation of new safety and environmental regulations for projects under construction.”[103] WPPSS's answer to the rising construction costs was to push ahead ever more rapidly. BPA viewed the construction problems and cost increases at WPPSS with singular detachment. In BPA's annual report for 1976, the main concern seemed focused not on the troubled construction but on WPPSS financing.[104]

The Failure of WPPSS Managerial Reforms

The years 1977–80 saw a steady increase in the projected costs of the WPPSS nuclear plants. It was not until 1980, however, that the true nature of WPPSS predicament dawned on both its own management and BPA. By this time the situation was out of control. In June 1980 the BPA noted that the estimated completion costs for the five projects had risen from $11.8 billion to $15.9 billion. WPPSS was also in the grip of a second prolonged five-month labor shut-down which lasted from June to November 1980. This meant that in the period between ground-breaking in 1972 and the end of 1980 WPPSS lost nearly a year of construction time due to labor disputes. This was certainly the result of WPPSS's construction policy, which emphasized speed and rendered it susceptible to union blackmail. By 1980 WPPSS bonds for units #4 and #5 bore an interest rate of 12.44 percent. In early 1980 WPPSS appointed a new managing director, Robert L. Ferguson, who at the time headed the U.S. Department of Energy's Nuclear Reactor Program. Ferguson had previously been in charge of the construction of the Hanford Fast Flux Test Facility and had managed to bring it in on schedule and at budget cost. Ferguson immediately instituted a drastic reorganization of the WPPSS management. Simultaneously BPA, "with the full concurrence of Bob Ferguson and his staff," strengthened its oversight of the WPPSS projects by sending to WPPSS six professionals including a Hanford resident manager "officed with the WPPSS top management team." In addition BPA's budget and audit offices began to work closely with their WPPSS counterparts.[105]

Robert Ferguson brought to WPPSS a professional approach to management. He made quick and radical changes to WPPSS's policy. Under previous administrations WPPSS had acted much like the political entity that it was. Controlled by about twenty different municipalities and PUDs, WPPSS treated the projects as a bonanza and spread the contracts around. Whereas most nuclear plants had one prime contractor, WPPSS had between forty-five and sixty at each unit. WPPSS itself attempted to supervise these contractors. Unlike the vertically integrated power systems such as TVA, WPPSS had no experienced staff when construction began but had to build it as the projects expanded. It is little wonder that the organization lost control of what turned out to be one of the largest nuclear plant construction programs in United States history (only the TVA's program was larger). Ferguson immediately moved to correct this situation. He went to the Washington State Legislature and sought a law changing the way the System could enter into contracts. The new act removed the requirement that the System accept the lowest bid and allowed it to make a decision based upon the contractor's ability to carry out the work. The labor conflict that had brought construction at the Hanford site to a halt was resolved. In February 1980 WPPSS negotiated a labor stabilization agreement at Satsop that provided for arbitration, and eliminated both picket lines and lockouts.

He also moved toward similar arrangements at Hanford. Ferguson also worked closely with the NRC to end its stop-work order on project #2, which had been issued in July 1980. Because of defective construction on the nuclear shield, the NRC required a thorough investigation and the reverification of all documentation to ensure that standards of quality had actually been met.[106]

For the first time, WPPSS projects began to move forward at scheduled speeds and within budget. One of Ferguson's major tasks had been to make reliable cost estimates. As can be seen from an examination of table 1, WPPSS never achieved solid estimates of what its projects would cost. WPPSS merely revised the figures every year. Its administrators apparently thought WPPSS could float unlimited bonds. When Ferguson became managing director this perception changed. His goal was to restore confidence in WPPSS by efficient contracting and building procedures and by developing reliable cost data. Despite Ferguson's successes his efforts were too little and too late. In January 1981 he began a detailed costing review. At that time, WPPSS had estimated that its projects would cost approximately $15.5 billion. Ferguson's analysis raised the estimate to $23.78 billion. Ironically, Ferguson's realistic approach proved WPPSS's downfall. Wall Street reacted in horror to the new figures. Given WPPSS's record, no one had any confidence in the statistics. WPPSS's then-current bond issues, which totalled over $9.3 billion, had glutted the market. Wall Street could not absorb another $14.4 billion. No amount of good management could change this fact.

Problems in Power Consumption Forecasting

At the root of WPPSS's crisis were the incorrect forecasts of Pacific Northwest power demand. In all fairness, it must be acknowledged that WPPSS did not generate these forecasts: the most influential element in the power forecasting was BPA. BPA's willingness to underwrite three WPPSS plants through net billing, together with its July 1976 notices of insufficiency to its preference customers, underpinned the WPPSS expansion program. WPPSS would not have been able to find participants for units #4 and #5 had not BPA issued its insufficiency notice when it did. How could BPA have been so wrong?

At the time, WPPSS was trying to sign up participants for its units #4 and #5, power forecasting in the Pacific Northwest had become very controversial. Both environmentalists and the proponents of low-cost power questioned the need for a massive investment in thermal generation. Both groups favored conservation. A reduction in power consumption not only avoided the need for new potentially pollution-generating capacity but it also ensured that most users would retain access to low-cost hydropower provided by existing federal facilities. Controversy flared in 1975 when Seattle City Light authorities suggested that the municipality purchase 10 percent of WPPSS projects #4 and #5. Environmentalists threatened to prevent this by legal action. The city, which had merely taken an option to participate in the projects pending an investigation of power needs, agreed to include environmental factors in the study. The investigation, undertaken by the Northwest Environmental Technical Laboratories, produced its report (*Energy 1990*) in 1976. The authors of *Energy 1990* assumed that power usage would decrease as power prices increased. Because BPA rates were rising, they forecast markedly lower energy requirements for Seattle and their outlook was quickly validated when power demand in the year 1978–79 fell below even their own modest projections.[107] As a result of the *Energy 1990* survey Seattle elected not to participate in WPPSS projects #4 and #5. In 1977 Roger Beers and Terry Lash, working under the auspices of the National Resources Defense Council, issued a report (*Choosing an Electrical Future for the Pacific Northwest: An Alternative Scenario*) which supported the conclusions reached in the Seattle study.[108] BPA, in its 1976 report, admitted that "electric utility forecasts are subject to increasing skepticism . . . " and bemoaned the fact that "nearly every proposed new power plant is seemingly being challenged on the basis that its output may not be needed." BPA mentioned the Seattle 1990 report and opined that "until some accepted mechanism can be developed for validating electric energy forecasts, power supply planning will be plagued by disputes. . . . "[109] Nevertheless, BPA was not swayed; it reacted by publishing its notices of insufficiency. The following year, 1977, BPA again lamented the controversy. While admitting that "few energy subjects are as contentious as load forecasting," it stated that "even the most conservative estimates of Northwest power requirements in the early 1980s are likely to be unmet with critical water conditions."[110]

In hindsight BPA's attitude is hard to defend. The Pacific Northwest's energy consumption growth rate started to decline even before WPPSS began planning its first nuclear plant in 1970. Furthermore the growth rate kept declining and fell continuously below the mainstream power forecasting estimates. Yet from an institutional viewpoint BPA's behavior is thoroughly understandable. Thomas P. Hughes, in *Networks of Power*, analyzes the growth of power "systems" in the first three decades of the twentieth century. Once a system solves its critical problems, he observes, it "usually grows if there is an adequate demand for its product." The system develops a direction and goals and as it "grows, it acquires momentum." The organization also develops a "culture" which fosters this growth.[111]

BPA developed its culture during the New Deal's fight for public power. New Deal public power advocates claimed that high rates charged by IOUs had retarded electricity use. The New Dealers believed in economies of scale in electricity production, arguing that most utilities had excess capacity and would actually generate more revenue by lowering charges. They maintained that power usage was highly elastic and that consumers would immediately respond to lower rates. In the early 1930s, average home consumption of electricity in the United States was 595 kilowatt hours. TVA authorities felt that low rates could double home consumption to about 1,200 kilowatt hours. In 1933 one of TVA's directors, David Lilienthal, announced dramatic rate cuts to communities served by the power authority. For example, electricity charges at the Alcorn County Cooperative declined from 5.37¢ per kilowatt hour to 1.8¢, while usage rose from 49 kilowatt hours per month to 139. A similar rate decline in Athens, Alabama tripled the demand for electricity.[112] TVA's philosophy forced Wendell Willkie's Commonwealth & Southern to experiment with low rates. The experience of Commonwealth

& Southern validated the contention of public power advocates that low rates would increase consumption.[113] At TVA and later BPA the mission became the encouragement of power use through low charges. Public power succeeded admirably. BPA set its rates in 1937 and managed to go until 1965 before raising them and then the increase was only 3 percent! This rate structure was maintained for another nine years before rates were increased 27 percent in 1974.[114] Because the BPA kept prices stable during a period of increasing inflation, the real cost of Pacific Northwest power declined dramatically until 1974. The region's power usage indicated this. While the typical Pacific Northwest household used nearly the same amount of energy as the national average, it used nearly twice as much electricity and paid about half as much for it.[115]

The New Deal aim to encourage the use of electricity was a good idea for the 1930s and 1940s. Those were years when many farms had still to be connected to the electric grid, and the extensive use of electric appliances had only begun. There was still ample hydropower to be developed in many parts of the nation, particularly the Northwest, and other energy supplies were abundant.

But energy policies of the 1930s were unsatisfactory for the 1970s. By that decade power sources were not abundant, but scarce. Electricity use had become universal and, more important, neither prosperity nor the good life was necessarily viewed as dependent upon the continuing expansion of electric power consumption. Furious opposition to the hydroelectric development of the middle reach of the Snake River indicated that public opinion was changing. The proposed Middle Snake Hydropower Project should seemingly have been irresistible: for a mere $300 million two dams could be constructed with a power rating of 2,700 megawatts. This was more than twice the amount that could be produced at one of WPPSS nuclear stations and it was estimated that the project could be completed in half the time. Despite these overwhelming advantages the "wild river" advocates triumphed over the developmentalists and the middle reach of the Snake remained free of dams.

Nevertheless BPA administrators continued to believe in the New Deal dream of cheap power. Their future predictions were based largely on historic trends. These in turn were heavily influenced by BPA's rate policy, which caused the real cost of power to decline continuously until 1974. BPA's executives forgot the New Deal experience which demonstrated that power demand was highly price-sensitive. The Pacific Northwest power demand was undoubtedly more cost-sensitive than that in most areas because of the importance of the region's large industrial users, especially the aluminum industry. Many of the aluminum mills were aging and approaching the time when they needed rebuilding. By the 1960s the aluminum companies were all multinationals with smelting operations scattered around the world in countries such as Canada, Jamaica, and Australia. Price rises to these firms quickly forced the transfer of aluminum smelting from the Pacific Northwest to more favorable areas. This automatically increased Columbia River power supplies available to the general public by as much as 15 percent or more. The rate projections which used the econometric methods employed by the consultants for Seattle and by the Natural Resources Defense Council clearly comprehended the relationship between price and power consumption. Blinded by the low-cost power philosophy of the New Deal, BPA did not use these methods—and failed. Nevertheless, it will be recalled that BPA was forced to raise power rates sharply, starting with an average 27 percent increase in 1974. Thereafter, BPA's rate increases ran firmly ahead of the nation's inflation rate. The results again validated what the New Dealers learned in the 1930s. It was not until the early 1980s that BPA's power projection model adopted the more sophisticated methodology which the Seattle consultants and the alternative power scenario advocates had pioneered. By this time it was too late for WPPSS.

Conclusions

In retrospect, the WPPSS fiasco has few villains. In the Tennessee Valley the New Deal created an authority that was a fully vertically integrated electric utility. From the start, the TVA gained experience in the construction of large-scale electric generating facilities. TVA also developed substantial transmission networks and its own retail distribution systems. Most important, TVA built a large and talented management team. It is true that TVA was also captivated by the New Deal concepts of cheap power and expanding electricity use. The organization undertook the largest nuclear power construction program in the United States. Yet the organization's depth of management and construction traditions enabled it to bring units on-line at fairly reasonable costs. Furthermore TVA's institutional structure allowed it to abandon or delay plants when it finally recognized that its projections, like those of BPA, were not accurate. TVA, in common with its private investor–owned counterparts, had a large cash flow from established projects. It also had the ability to raise rates so that it was able to meet its interest and debt obligations.

In contrast, the New Deal era produced a hopelessly fragmented public power system in the Pacific Northwest. Of all of the Northwest's public agencies, only the City of Seattle seemed capable of coping with the changing power requirements of the 1970s. Seattle's public system, which predated federal hydropower development, was a large utility which in most ways was comparable to its investor-owned neighbors such as Portland General Electric. Seattle had the financial strength to build its own generating capacity. Seattle Light's tradition gave it the confidence to act independently. In addition, the Seattle City Council, which controlled the municipal utility, was immediately sensitive to changing public perceptions. Seattle, a university city, financial center, and home of high-tech manufacturing, had an educated elite, who were quick to embrace environmental protection and energy conservation. Seattle's elite forced the city to reevaluate its power policies. After much controversy, the City Council acted to remove Seattle from WPPSS Projects #4 and #5.

In contrast, neither BPA nor WPPSS proved institutionally able to respond to new challenges. BPA was a federal entity, at first part of the United States Department of the Interior and later under the control of the Department of Energy. Its administrators were appointed by federal politicians who were remote to the Pacific Northwest. The agency itself had been designed in such a way that it had little power to initiate or control development in the Columbia River basin. During the time when WPPSS made its decisions to build nuclear plants, the Republicans, with first Richard Nixon and then Gerald Ford, controlled the Presidency. While the Republicans opposed the expansion of federal power, they favored nuclear power, especially since it could be substituted for power generated by expensive imported oil. BPA's political overseers had little inclination to interfere in the administration's relationship with WPPSS.

Of all of the institutions, WPPSS itself was the most flawed. While it was the creature of the State of Washington, its effective controllers were the eighteen or so public utility districts and the three municipalities which provided its directors. WPPSS was so constructed as to give the weight of power not to the experienced large municipal public utilities of Seattle and Tacoma but to the state's small rural PUDs. Initially these districts shared most of the same values that activated the New Deal public power program. Their ratepayers were accustomed to low-cost power and saw nothing wrong with continued expansion of the use of public power. These districts lacked the urban elite which spearheaded Seattle's rejection of WPPSS units #4 and #5. WPPSS directors, satisfied with BPA's performance in the past, were only too glad to take its advice about the shape of the Pacific Northwest's power future. Worse yet, the State of Washington had created WPPSS along the lines of a traditional municipal council but had given it no taxing power. Washington's legislators also encumbered WPPSS with administrative procedures which were totally unable to cope with the needs of a large technical challenge such as the building of nuclear plants. Given WPPSS's institutional constraints and the fragmentation of authority in the Pacific Northwest, it is not surprising that the experiment ended in failure.

NOTES

1. NUKEM 1988, pp. 39–41.

2. For exact figures see tables 1 and 2.

3. BPA, *1988 Financial Summary*, pp. 46, 48. Data on the assets of Southern, CSX, and UAL in 1980 dollars: Pillsbury 1981, pp. 124, 126.

4. *Wall Street Journal* (hereafter *WSJ*), 22 June 1981.

5. For the background to the Public Power Preference Clause see McCraw 1971, p. 53.

6. BPA, *Annual Report, 1976*, pp. 22–23.

7. See table 3.

8. Roosevelt 1938, p. 728.

9. For the difference in rates between government and private corporate borrowing see table 3.

10. The government itself received no return for navigation, flood control, etc.

11. BPA, *1988 Financial Summary*, pp. 46–47.

12. Data from *United States Code: Congressional and Administrative News, 96th Congress—2nd Session, 1980*, vol. 5: *Legislative History, Public Laws 96-472 to 96-522* (St. Paul, Minnesota: West Publishing Co.; Pacific Northwest Electric Power Planning and Conservation Act), p. 6019.

13. *Ibid*., p. 5994.

14. BPA, *Annual Report, 1977*, p. 13.

15. BPA, *Annual Report, 1976*, p. 20.

16. Roosevelt 1938, p. 738.

17. McCraw 1971, p. vii.

18. Republican Senator Bourke Hickenlooper as quoted in Lee and Klemka 1980, p. 62. This description of the federal government's building of the N-reactor and the beginning of WPPSS is drawn from Lee and Klemka's splendid book on the Pacific Northwest's electric power problems. Their work is clearly the best published account of this thorny issue and my analysis of WPPSS has been greatly aided by their pathbreaking study.

19. WPPSS, *Background Information, October, 1987*, p. 1. See also Lee and Klemka 1980, p. 23.

20. Lee and Klemka, *Electric Power*, pp. 64–65.

21. NUKEM 1988, p. 39.

22. Power costs: WPPSS, *Annual Report, 1973*, p. 13, and BPA, *Annual Report, 1976*, p. 14.

23. For a discussion of the Pacific Northwest's Hydrothermal Plan, see Lee and Klemka 1980, pp. 66ff.

24. See table 4.

25. See table 3.

26. BPA, *Annual Report, 1977*, p. 33.

27. For data on net billing see WPPSS, *Annual Report, 1973*, pp. 14–17.

28. This and other quotations from BPA, *Annual Report, 1976*, pp. 4–5.

29. BPA, *Annual Report, 1976*, pp. 1–2.

30. BPA, *Annual Report, 1976*, p. 3.

31. *Ibid*., p. 5.

32. WPPSS, *Annual Report, 1976*, p. 28.

33. BPA, *Annual Report, 1977*, p. 7.

34. WPPSS, *Annual Report, 1973*, p. 14.

35. NUKEM 1988, p. 39.

36. Science Policy Research Division Congressional Research Service 1973, p.3.

37. Goodwin 1981, pp. 410, 445–54.

38. Dorf 1981, p. x.

39. WPPSS, *Annual Report, 1975*, p. 17.

40. See table 4 for data on interest rates from 1937 to 1982.

41. WPPSS, *Annual Report, 1973*, p. 6.

42. WPPSS, *Annual Report, 1978*, p. 10.

43. BPA, *1980 Program and Financial Summary*, p. 22.

44. BPA, *1983 Program and Financial Summary*, p. 19.

45. See table 4.

46. Montana Power Company, *Annual Report, 1988*, p. 29; Montana Power Company, "Interim Letter to Shareholders," 30 June 1989.

47. WPPSS, *Annual Report, 1975*, p. 12.

48. Public Law 96-501, 5 December 1980.

49. *United States Code: Congressional and Administrative News, 96th Congress, 2nd sess., 1980*, vol. 5: *Legislative History*, p. 5992.

50. BPA, *1980 Program and Financial Summary*, pp. 25–27.

51. *WSJ*, 29 June 1981.

52. For analysis see *Business Week*, 11 July 1983, p. 43.

53. *WSJ*, 24 July 1981.

54. *Ibid.*

55. *WSJ*, 25 September 1981.

56. *WSJ*, 25 January 1982.

57. Cincinnati Gas & Electric Company, *Annual Report 1988*, p. 26. See also Cincinnati Gas & Electric Company, *Company Financial and Statistical Review, 1978–1988*, p. 5.

58. *WSJ*, 25 January 1982.

59. BPA, *Annual Report, 1981*, p. 7; BPA, *Annual Report, 1982*, p. 14; BPA, *Program and Financial Summary, 1983*, p. 5.

60. *WSJ*, 5 November 1981.

61. Tamietti 1984, p. 278.

62. *Ibid.*, p. 278–89.

63. WPPSS, *Annual Report, 1985*, p. 26.

64. *Business Week*, 30 May 1983, pp. 22–23.

65. Brimelow 1983, p. 46.

66. *Ibid.*, p. 47.

67. For example of this see WPPSS, *Annual Report, 1985*, pp. 30–31.

68. WPPSS, a letter entitled "Issue Paper, Refinancing High-interest Debt" (February 1989).

69. Hughes 1983, p. 7.

70. Washington Water Power Company, *Annual Report, 1988*, pp. 1–5.

71. For a full discussion of the growth of this system see Hughes 1983, pp. 324–34.

72. For the best short analysis of this fight see McCraw 1971. This volume has a lucid analysis of Willkie's strategy to limit the danger of TVA to IOUs.

73. *Ibid.*, ch. 7.

74. For a clear, concise discussion of the rise of BPA see Blumm 1982–83, esp. pp. 191–209.

75. *Ibid.*, 209–10.

76. WPPSS, *Annual Report, 1977*, p. 4.

77. The number varied but normally was between seventeen and twenty-one.

78. WPPSS, *Annual Report, 1973*, p. 9; WPPSS, *Annual Report, 1978*, p. 7.

79. NUKEM 1988, p. 39.

80. Bischak 1988, p. 70.

81. *WSJ*, 25 March 1975.

82. *New York Times*, 22 September 1975.

83. *WSJ*, 24 February 1986.

84. Shapley 1977, pp. 1311–13.

85. *Ibid.*, p. 1313.

86. Passer 1953, pp. 268–70.

87. See comments by Bernstein (1982, p. 106).

88. Data compiled from annual reports of the Tennessee Valley Authority, and *World Nuclear Industry Handbook (1987)*, pp. 171–72.

89. See table 1.

90. See table 2.

91. See table 2.

92. WPPSS, *Annual Report, 1974*, p. 10.

93. *Ibid.*, p. 7.

94. See WPPSS, *Annual Report, 1976*, p. 30.

95. See table 2.

96. See table 1.

97. *New York Times*, 26 March 1975.

98. WPPSS, *Annual Report, 1975*, p. 7.

99. WPPSS, *Annual Report, 1975*, p. 17.

100. *WSJ*, 7 September 1983.

101. WPPSS, *Annual Report, 1980*, p. 13.

102. For a description of the accounting procedures which WPPSS used see WPPSS, *Annual Report, 1984*, p. 23, note B.

103. WPPSS, *Annual Report, 1975*, p. 7.

104. BPA, *Annual Report, 1976*, pp. 4–5.

105. BPA, *1980 Program and Financial Summary*, pp. 13–14.

106. Data for this paragraph comes from WPPSS, *Annual Report, 1981* (various pages—report

unpaginated), and Bernstein 1982, pp. 106, 110.

107. The best summary of *Energy 1990* and the Seattle debate is found in Lee and Klemka 1980, pp. 122–29.

108. Palo Alto: National Resources Defense Council, January 1977.

109. BPA, *Annual Report, 1976*, p. 2.

110. BPA, *Annual Report, 1977*, p. 4.

111. Hughes *1983*, p. 15.

112. McCraw 1971, pp. 60–61, 74.

113. *Ibid.*, pp. 74–77.

114. BPA, *Annual Report, 1977*, p. 7.

115. WPPSS, *Annual Report, 1977*, p. 9.

116. WPPSS, *Annual Report, 1973*, p. 19.

REFERENCES

Bernstein, Peter W. 1982. "A Nuclear Fiasco Shakes the Bond Market." *Fortune* (22 February), p. 106.

Bischak, Gregory Allen. 1988. *State-Induced Technological Change in the United States Nuclear Power Industry, 1947–1987*. Ann Arbor: UMI.

Blumm, Michael C. 1982–83. "The Northwest's Hydroelectric Heritage: Prologue to the Pacific Northwest Electric Power Planning and Conservation Act." *Washington Law Review* 58.

Brimelow, Peter. 1983. "Shock Waves from Whoops Roll East." *Fortune* (25 July), p. 46.

Dorf, Richard C. 1981. *The Energy Fact Book*. New York: McGraw Hill.

Goodwin, Craufurd D. 1981. *Energy Policy in Perspective: Today's Problems, Yesterday's Solutions*. Washington, D.C.: Brookings.

Hughes, Thomas P. 1983. *Networks of Power: Electrification in Western Society, 1880–1930*. Baltimore: Johns Hopkins University Press.

Lee, Kai N., and Klemka, Donna Lee (with Marion E. Marts). 1980. *Electric Power and the Future of the Pacific Northwest*. Seattle: University of Washington Press.

McCraw, Thomas K. 1971. *TVA and the Public Power Fight 1933–1939*. Philadelphia: J. P. Lippincott.

NUKEM. 1988. *Special Report* 3.

Passer, Harold. 1953. *The Electrical Manufacturers: 1875–1900*. Cambridge: Harvard University Press.

Pillsbury, Anne B. 1981. "The *Fortune* Directory of the Largest Non-Industrial Companies: The 50s." *Fortune* (13 July), pp. 124, 126.

Roosevelt, Franklin D. 1938. *The Public Papers and Addresses of Franklin D. Roosevelt.* Vol. 1. New York: Random House.

Shapley, Deborah. 1977. "Nuclear Power Plants: Why Do Some Work Better Than Others?" *Science* (25 March), pp. 1311–13.

Science Policy Research Division Congressional Research Service. 1973. *Energy Facts,* Prepared for the U.S. House of Representatives Subcommittee on Energy of the Committee of Science and Astronautics, 93rd Congress, 1st sess. Library of Congress Serial H (November).

Tamietti, Robert L. 1984. "Chemical Bank v. WPPSS: A Case of Judicial Melt-down." *Journal of Energy Law and Policy* 5: 278.

TABLE 2

Washington Public Power Supply System: Expenditure for Nuclear Plants*

	WPPSS #2[†][a]	WPPSS #1[b]	WPPSS #3[c]	WPPSS #4[d]	WPPSS #5[e]
31 December 1973					
Expended	$ 65,618,000	$ 8,292,100	$ 2,559,200[f]		
Committed by WPPSS	156,363,000	85,796,000	190,100,000[g]		
31 December 1974					
Total expended to date	137,883,000	17,555,000	10,759,000	n.a.	n.a.
Committed by WPPSS	272,648,200	127,947,400	102,657,100[j]	$ 116,978,300	$ 117,197,600
30 June 1975					
Total expended to date[k]	191,828,000	27,348,000	18,301,000	8,008,000	
Committed by WPPSS[k]	280,000,000	186,000,000	227,000,000	364,000,000	
30 June 1976					
Total expended to date	373,134,000	115,431,000	38,078,000[f]	29,605,956	8,328,282
Committed by WPPSS[k]	186,000,000	216,000,000	289,000,000[f]	372,000,000	
30 June 1977					
Total expended to date	509,378,000	172,527,000	81,543,000[f]	112,957,000	50,744,000[f]
Committed by WPPSS	183,000,000	260,000,000	299,000,000	296,000,000	270,000,000
30 June 1978					
Total expended to date	736,587,000	365,017,000	147,963,000[f]	276,819,000	179,546,000
Committed by WPPSS	n.a.	n.a.	n.a.	n.a.	n.a.
30 June 1979					
Total expended to date	948,000,000	629,037,000	252,439,000[f]	416,806,000	307,323,000[f]
Committed by WPPSS	n.a.	n.a.	n.a.	n.a.	n.a.
30 June 1980					
Total Expended to date	1,111,470,000	930,634,000	440,182,000[f]	666,469,000	495,649,000[f]
Committed by WPPSS	n.a.	n.a.	n.a.	n.a.	n.a.
30 June 1981					
Total expended to date	1,291,646,000	1,207,000,000	669,019,000	951,000,000	821,317,000[f]
Committed by WPPSS	n.a.	n.a.	n.a.	n.a.	n.a.
30 June 1982					
Total expended to date[k]	2,133,978,000	2,034,000,000	1,106,214,000	2,241,482,000	
Total long-term WPPSS debt[k]	3,329,870,000	2,151,305,000	1,600,000,000	2,250,000,000	
30 June 1983					
Total expenditures, plant and equipment[k]	2,750,514,000	2,265,092,000	1,547,122,000	2,424,172,000[l]	
Total long-term debt[k]	2,314,800,000[m]	2,147,490,000[m]	1,600,000,000	2,317,866,000[n]	

Source: WPPSS Annual Reports, years 1973–88.

*Figures for total projects unless otherwise indicated.

[†]Nuclear plants listed in order of construction.

[a]Planned 1970; ground-breaking took place in August 1972.

[b]Approved February 1973.

[c]Approved June 1973.

[d]Approved July 1974.

[e]Approved July 1974. Original ownership to be 70 percent WPPSS and 30 percent investor-owned utilities: Pacific Power & Light 10 percent, Portland General Electric 10 percent, Puget Sound Power & Light 5 percent, Washington Water Power 5 percent. All but Pacific Power refused to join, and ownership as of 1976 was 90 percent WPPSS and 10 percent Pacific Power.

[f]WPPSS portion of project only; 30 percent owned by investor-owned utilities (Pacific Power & Light 10 percent, Portland General Electric 10 percent, Puget Sound, Power & Light 5 percent, and Washington Water Power 5 percent).

[g]Long-term cost of rods for reactor, fuel, etc.

[h]WPPSS, *Annual Report 1974*, p. 24.

[i]WPPSS, *Annual Report 1978*, p. 31.

[j]Total for both WPPSS share and other participating companies.

[k]Shows combined figure for #4 and #5.

[l]Costs related to construction and termination of utility plants.

[m]Some debt retired through cash payments from Bonneville Power Administration as a result of net billing.

[n]Debt in default currently payable. In addition there were $67,865,000 of subordinated revenue notes in default. Note as of 1988 this debt is still in default with no hope of payment.

TABLE 3

Interest Rates 1937–82

(percent per annum)

Type of Security	1937	1938	1940	1945	1946	1947	1948	1949	1950
U.S Government	2.74	2.61	2.26	2.37	2.19	2.25	2.44	2.31	2.32
20-year Treasury Bonds[a]	- -	- -	- -	-0.-	- -	- -	- -	- -	- -
High-grade Municipal Bonds	3.10	2.91	2.50	1.67	1.64	2.01	2.40	2.21	1.98
Corporate	3.26	3.19	2.84	2.62	2.53	2.61	2.82	2.66	2.62

Type of Security	1951	1952	1953	1954	1955	1956	1957	1958	1959	1960	1961	1962	1963	1964	1965
U.S. Government	2.57	2.68	2.94	2.55	2.84	3.08	3.47	3.43	4.07	4.01	3.90	3.95	4.00	4.15	4.21
20-year Treasury Bonds[a]	- -	- -	- -	- 5-	- -	- -	- -	- -	- -	- -	- -	- -	- -	- -	- -
High-grade Municipal Bonds	2.00	2.19	2.72	2.37	2.53	2.93	3.60	3.56	3.95	3.73	3.46	3.18T3.23	3.22	3.27	
Corporate	2.86	2.93	3.20	2.90	3.06	3.36	3.89	3.79	3.38	4.41	4.35	4.33	4.26	4.40	4.49
WPPSS Hanford Nuclear Plant and Packwood Lake												3.66	3.26	3.76	

Types of Security	1966	1967	1968	1969	1970	1971	1972	1973	1974	1975	1976	1977	1978	1979	1980	1981	1982
U.S. Government	4.66	4.85	5.25	6.10													
20-year Treasury Bonds[a]	- -	- -	- -	- -T6.86	6.12	6.01	7.12	8.05	8.19	7.86	7.67	8.48	8.74	10.81	12.57		12.23
High-grade Municipal Bonds	3.82	3.98	4.51	5.81	6.51	5.70	5.27	5.18	6.09	6.89	6.49	5.56	5.90	6.39	8.51	11.23	11.57
Corporate (Moody's)	5.13	5.51	6.18	7.03	8.51	7.94	7.63	7.80	9.03	9.57	9.01	8.43	9.07	10.12	12.75	15.06	14.94
WPPSS #1 and #2								5.66	7.21	6.71	6.63		6.71	6.49	9.36	12.44	14.76
									7.67		5.87			7.69			13.82
										7.87	6.48	5.71	6.27			10.80	14.83
																14.80	13.95
																	13.63
WPPSS #4 and #5												5.93	6.07	7.15	9.23	11.77	
												6.32	6.86	7.69	9.50	11.06	
												5.96	6.81	8.30	10.69		
															12.44		
															11.83		

Source: Through 1970: U. S. Bureau of the Census, *Historical Statistics of the United States, Colonial Times to 1970* (Washington D.C., 1975), p. 1001; after 1970: *Statistical Abstract of the United States*, various years; WPPSS rates: *WPPSS Annual Report*, various years ("effective interest rate").

[a]After 1979 "Long-term" U.S. Government bonds.

TABLE 4

West Group (Pacific Northwest): Hydrothermal Power Program as of 1978

PHASE 1

Project Number	Principal Sponsor	Location	Type	Rated Capacity (mw)	Date of Commercial Operation
1	Pacific Power & Light Co. & Washington Water Power	Centralia, Washington	coal-fired	1,400	1971
2	Portland General Electric Co. (Trojan Project)	near St. Helens, Oregon	nuclear	1,130	1976
3	Pacific Power & Light Co. (Jim Bridger Project)	Rock Springs, Wyoming	coal-fired	500 500	1975 1976
4	Montana Power and Puget Sound Power & Light Co.	Colstrip 1 & 2	coal-fired	700	In operation as of 1985
5	Washington Public Power Supply System (#2)	Hanford, Washington	nuclear	1,100	1984
6	Portland General Electric Co. (Pebble Springs #1)	Arlington, Oregon	nuclear	1,200	Cancelled
7	Washington Public Power Supply System (#1)	Hanford, Washington	nuclear	1,206	Delayed indefinitely
8	Washington Public Power Supply System (#3)	Satsop, Washington	nuclear	1,240	Delayed
9	Portland General Electric	Boardman, Oregon	nuclear	1,260	Not built, environmental objections

PHASE 2

Project Number	Principal Sponsor	Location	Type	Rated Capacity (mw)	Date of Commercial Operation
1	Pacific Power & Light Co. (Jim Bridger #4)	Rock Spring, Wyoming	coal-fired	500	In operation as of 1985
2	Pacific Power & Light & Washington Water Power (Centralia #3)	Centralia, Washington	coal-fired	700	In operation as of 1985
3	Portland General Electric Boardman Coal (Carty)	Boardman, Oregon	coal-fired	530[a]	In operation as of 1985
4	Washington Public Power Supply System (Hanford N-reactor continued beyond 1977)	Hanford, Washington	nuclear	860	N-reactor shutdown for safety reasons; restarting uncertain
5	Puget Sound Power & Light Company (Skagit #1)	Sedro Woolley, Washington	nuclear	1,288	Cancelled
6	Puget Sound Power & Light Company (Colstrip #3)	Colstrip, Montana	coal-fired	700[b]	In operation as of 1985
7	Puget Sound Power & Light Company (Colstrip #4)	Colstrip, Montana	coal-fired	700[b]	April 1986
8	Washington Public Power Supply System (Thermal Project #4)	Hanford, Washington	nuclear	1,240	Cancelled
9	Portland General Electric (Pebble Springs #2)	Arlington, Oregon	nuclear	1,260	Cancelled
10	Puget Sound Power & Light Co. (Skagit #2)	Sedro Woolley, Washington	nuclear	1,288	Cancelled
11	Washington Public Power Supply System (#5)	Satsop, Washington	nuclear	1,240	Cancelled

Sources: WPPSS, *Annual Report, 1973*; BPA, *Annual Report, 1978*; Power Planning Committee Pacific Northwest Business Commission, *Review of Power Planning in the Pacific Northwest, Calendar Year 1973* (printed May 1974), pp. 4, 31; Lee and Klemka, *Electric Power*, p. 134; Electric World, *Director of Electric Utilities 1985–1986.*, 94th ed. (New York: McGraw Hill).

[a]90 percent used by West Group.

[b]60 percent used by West Group.

IRAN AIR FLIGHT 655
AND THE USS *VINCENNES*

Complex, Large-Scale Military Systems and the Failure of Control

Gene I. Rochlin
Institute of Governmental Studies and
Energy and Resources Group
University of California at Berkeley

Preface: U.S.S. *Vincennes* and Iran Air Flight 655

Patrolling the restricted waters of the Persian Gulf was a trying activity for most U.S. warships, designed, armed, and trained as they were for far-ranging "blue water" operations. This was particularly true for the officers and crew of the USS *Vincennes*. One of the first of the *Ticonderoga*-class "Aegis" cruisers, the *Vincennes* is a fast, lightly armored ship—a cruiser built on a large destroyer hull—specially optimized for fleet air defense. Although armed with various surface-to-surface guns and a variety of systems for close-in air defense, her real "main battery" consisted of the Standard SM-2 anti-aircraft missiles stored deep in her magazines.

In her normal mission of providing air defense to an aircraft carrier battle group, the *Vincennes*'s advanced Aegis fire-control system was capable of projecting a visible image of an air battle of many hundred square miles, tracking and distinguishing friendly and potentially hostile aircraft at ranges of tens of miles while engaging a variety of potential targets ranging from high-flying reconnaissance aircraft to high-speed cruise missiles. Bottled up in confined waters, this billion-dollar bundle of sophisticated and advanced technology was not much more able to defend herself from mines and Iranian speedboats than a destroyer, and was almost as vulnerable.

But the U.S. Navy, with its focus on broad-ocean task forces and quasi-strategic "maritime strategies," had not built a coastal patrol navy, relying on its European allies to perform this function for them in NATO waters. So, on the morning of 3 July 1988, the *Vincennes*, like many of the U.S. ships on Persian Gulf patrol, was engaged in a sweep of the shallow waters of the Straits of Hormuz, a mission for which traditional Navy skills such as ship-handling and gunfire were more important than the high technology aboard. Also in the vicinity were two U.S. frigates, the USS *Elmer Montgomery* (FF 1082) and the USS *Sides* (FFG 14). With the memory of the attack on the USS *Stark* the previous year still fresh in every sailor's mind, all aircraft detection and warning systems were up and fully manned.[1]

On the previous day, several armed small boats of the Iranian Revolutionary Guard Corps (IRGC) had positioned themselves at the western approach to the Straits, and were challenging merchant vessels. Late that day, the *Montgomery* had come close enough to a ship attack in progress to fire warning shots at several of the IRGC boats.

T. R. La Porte (ed.), Social Responses to Large Technical Systems, 99–125.

Early in the morning of 3 July the *Montgomery*, at the northern end of the Straits, reported an attack by seven small IRGC boats armed with machine guns and rockets. Shortly thereafter, another wave of thirteen such boats was reported, in three groups, one of which took a position at the *Montgomery*'s port quarter. At 7:42 A.M. local time, the *Vincennes* was dispatched to the area to investigate the situation. At about 9:45 A.M., one of the helicopters sent out to monitor the situation having been fired upon, the *Vincennes* went to General Quarters and took tactical command of the *Montgomery*, and both ships proceeded toward the incident at high speed.

As they approached the position of the boats, several were observed to turn toward the U.S. ships and close in a threatening manner. Taking this as *prima facie* evidence of hostile intent, Middle East Joint Task Force Command gave permission to engage, and the *Vincennes* opened fire at 10:13 A.M., starting a surface melee that was to continue throughout the incident. The IRGC boats, fully aware of the advantage conferred by their small size and maneuverability, did not flee, but turned to engage, hoping thereby to inflict some damage on the far more expensive and valuable U.S. ships.

At 10:16, the *Vincennes* opened fire with her 5" main guns. Four minutes later, a round that would not fire fouled the fore 5" mount, forcing the Tactical Operations Officer (TAO) to maneuver the ship radically—using 30 degrees of rudder at a ship's speed of 30 knots—to bring the aft 5" mount to bear on the most threatening of the small boats. This drastic, high-speed maneuver caused the ship to heel dramatically, sending loose equipment flying throughout the ship.[2] The Combat Information Center (CIC), which contains not only the Aegis displays but all other displays and consoles from which the ship is fought, was not spared; books, publications, and loose equipment went flying off desks, and many of those on duty had to grab for the nearest support to avoid following them to the deck. The surface engagement ended at 10:33, the *Vincennes* having expended seventy-two rounds of 5" ammunition and the *Montgomery* forty-seven. The IRGC boats were then in retreat, one of them having been sunk by U.S. gunfire.

That same morning, while the *Vincennes* and *Montgomery* were hightailing it toward the small boats north and west of the Straits of Hormuz, the captain and crew of Iran Air flight 655 were at Bandar Abbas airfield in southern Iran, preparing for the routine 150-mile flight over the Gulf to Abu Dhabi. It was one of the many ironies of the "half-war" in the Gulf that such commerce proceeded almost routinely, in, around, and through what amounted to open combat zones. Even the status of Bandar Abbas itself was ambiguous, as the newly arrived flight of Iranian F-14 fighters so clearly demonstrated.

Scheduled departure time was 9:59 A.M. local time, with the usual mixed load of business men and relatives making the quick hop to the comparative peace and luxury of the eastern Gulf. The flight was assigned routinely to commercial air corridor Amber 59, a twenty-mile-wide lane on a direct line to Dhubai airport. Owing to the short distance, the flight pattern would be more or less of a trajectory—climbing out to an altitude of 14,000 feet, cruising for a short time, then beginning a gradual descent into Dhubai.

After taking off from runway 21, Iran Air Flight 655 was directed by the Bandar

Abbas tower to squawk IFF Mode III, code 6760. As has been verified by the data from the USS *Sides*, the Captain did turn on his aircraft identification transponder. This unit is a modern form of the old "IFF" ("identification—friend or foe?") system of World War II. When interrogated by a radar signal from a potential adversary, the transponder "squawks" (gives off a specific response signal) in a prespecified and fixed mode. The Iranian F-14s at Bandar Abbas are presumed to have been set to squawk in "Mode II," a mode that would identify to the U.S. ships that the aircraft in question were Iranian, and military. Iran Air Flight 655, however, was set to squawk in Mode III, a signal that identifies a flight as civilian. The code number, 6760 in this case, would distinguish that particular flight from others.

In the case of Iran Air Flight 655, the matter of life or death seems to have been settled by the eighteen-minute delay in departure. Because of this timing, the flight first appeared on the *Vincennes*'s radar at 10:17, just after she had opened fire on the IRGC patrol boats. At 10:19, the *Vincennes* began to issue warnings on the Military Air Distress frequency, and at 10:20 to begin warnings on the International (civil) Air Distress frequency as well. It was at this moment, with the aircraft unidentified and apparently closing on the *Vincennes*, that the TAO ordered the radical maneuver that created disorder and tension throughout the CIC. Over the next three minutes, with the ship in a radical maneuver, the CIC in confusion and disorder, and while continuing to engage the IRGC boats, the *Vincennes* issued a number of warnings on both military and civil distress frequencies, (mistakenly) identified the Airbus 320 as a possible Iranian F-14, (mistakenly) reported hearing IFF squawks in Mode II, and (mistakenly) reported the aircraft as descending toward the ship when it was in fact still climbing according to its usual flight plan.[3]

Having informed Joint Task Force Command that a potentially hostile aircraft was at a distance of twenty-eight nautical miles and rapidly closing to within potential missile attack range, the *Vincennes* received permission to engage. Captain Rogers, the Commanding Officer (CO) held out for a minute or two more, by which time the still unidentified aircraft had closed to fifteen miles and was still being reported as descending toward his ship. At about 10:24 A.M., seven minutes into Iran Air Flight 655's flight, and eight minutes into *Vincennes*'s fire fight, the CO fired two SM-2 Standard missiles at the unknown target.

A few seconds later, with the flight still on its assigned climb-out, and slightly to one side of, but well within, air corridor Amber 59, it was intercepted by one or both of the missiles at a range of eight nautical miles and an altitude of 13,500 feet. The Airbus 300, with some 290 people from six nations aboard, tumbled in flames into the Persian Gulf. None survived. The whole flight had taken less than seven minutes.

By noon of that day, Iranian helicopters and boats began the search of the area and the recovery of the bodies. It was not until later that day that the officers and men of the *Vincennes* would learn that what they had shot down was not an Iranian F-14, but a commercial, civil flight.

Taken as an isolated event the case of the *Vincennes* may not seem to amount to much. In a tense situation in a combat zone, an unfortunate mistake caused the deaths of several hundred people. Iran was outraged, the U.S. embarrassed. Each accused the other of recklessness, of human and cognitive error. But I argue below that what happened over the Persian Gulf that day was illustrative of a whole new class of failures to which military large technical systems are increasingly susceptible as a consequence of the indirect organizational effects of the evolution of information and computer systems into large-scale networks of control. That the events of that day were not generally so analyzed, even by the several boards and panels convened to review them, points out the degree to which analysts of the military and analysts of complex, socio-technical systems have systematically neglected each others' field of research.[4]

Introduction: Modern Military Systems and Technologies of Control

Militaries have been large-scale systems at least since Roman times. Attempts have been made since time immemorial to increase the size, scope, technical sophistication, and complexity of military systems. Yet in every case historical constraints on command-and-control systems have imposed severe limitations—indeed have sometimes brought the most ambitious of plans to disaster.[5] Although similar limitations on what are now known as "C_3I" (command, control, communications, and intelligence) systems are still operative, the very sophistication of the communications and intelligence/information systems that added the third "C" and the "I" to "command and control" too often provides the illusion that events can be followed, interpreted, and controlled in real time. The illusion of control has serious implications for other large-scale technical systems that involve considerable potential risk.[6] In the case of military systems, particularly those that might involve nuclear weapons, the risk is incalculably greater.

The absence of adequate analysis of the military aspects of large-scale technological systems seems to arise at least partially from a historic, and quite normative, academic tradition of separating the study of the military from that of civil organizations, and leaving the former to specialists of the field—as if the military, and combat, were somehow outside of the normal range of political life. Military analysts in turn have traditionally tended to separate military studies into the study of battles, the study of men, and the study of machines. To the first category belong the studies of greater and lesser battles, both romantic and analytic—professional history and popular. To the second belong studies of "men in battle," again both romantic and professional—in modern terms, behavioral, psychological, and, occasionally, organizational studies. To the third we assign the increasingly numerous studies of military technology—again, often romantic, often historical, increasingly analytic, and occasionally both historical and organizational.[7]

It is the occasional appearance of the "organizational" in some of the more interesting recent work of military analysts and historians, particularly in the last category, that motivates this essay. Some have even gone so far as to note that military systems as large socio-technical systems are never that dissimilar in capabilities and structure from contemporary civil ones. Indeed, the whole category of "military technology" as separable from ordinary technology is a social construction of academic professionals, for the technological principles at work have much in common with those of contemporary civil technologies, however different the particularities of the military artifacts.[9]

Management vs. Control

The word "control" is used in many different ways, carrying with them many different meanings, even in the disciplinary literature.[10] Among the more precise is that put forward by Martin Landau to differentiate it from more traditional techniques of "management" according to the character and degree of knowledge involved.[11] In Landau's terms, any organization is fundamentally a seeker for knowledge—empirically verified observations, theories, and models not only of its own behavior, but of that of its organizational environment. But knowledge without models is completely "retrodictive." Since few decision circumstances ever recur, organizations seeking to act on the basis of knowledge seek comprehensive models, which attempt to predict future outcomes based on present information.

It is the nature and interpretation of models that makes the control-vs.-management distinction a particularly critical one for military systems. If one has perfect knowledge, correct information, and a verified, knowledge-based model that encompasses all possible variations, then one can indeed exercise "control" over outcomes. That is, one can take the information in, interpret it, fit it into the model, compare the modeled output with the desired output, and either accept or correct the action to keep events on the preordained course—all of which is predicated not only on the appropriateness of the selected model, but also on the congruence of the model's assumptions with evolving circumstances.

Management, on the other hand, involves decision-making under the acceptance of irreducible uncertainty, using heuristic models that are corrected on the fly, as necessary, as part of online trial and error learning. In life, there is always much more management than control. An outfielder drifting back under a fly ball in left field, a soccer goaltender watching the oncoming play, a batsman guarding a wicket, and a battlefield commander adjusting his troops according to the tide of battle are all "managing" their particular critical environment, adjusting on the fly to try and get a favorable outcome.

Technology and Control in the Military

The larger the scale and complexity of the military operation (e.g., the number of prospective combatants, the geographical scope, the diversity of task, and the nature and diversity of technology), the more difficult it is for the military commander to control events.[12] So long as information collection, analysis, and distribution systems were inadequate to the task of providing required data in real time, commanders had but few options: (a) to limit the size and scope of their armies so that all fell under more or less direct view; (b) to attempt to preshape events so as to bring them to a decisive point in time and space (so-called strategic battles); (c) to standardize units and tactics, and depend upon their autonomous action to bring about desired strategic outcomes; and (d) to preconstruct elaborate systemic designs, together with timetables and assignments, so that outcomes would be more or less assured if the plan was followed. Of these, (a), (b), and (c) all correspond more or less to various modalities of management. Of the historical modalities, only (d) corresponds to attempts at "control."

Each of the first three options can be documented by abundant historical evidence. Mode (a) was the choice of the highly professionalized armies of the eighteenth century. Napoleonic armies struggled for (b), and the Napoleonic record is a series of huge set battles. The Romans depended upon (c) for expanding and defending their extensive frontiers, especially to the north and west. That none of these was ever fully satisfactory to the commanders involved is also well documented, with the behavior of Moltke in the Franco-Prussian War perhaps the most famous of recent examples.[13] But the technological limitations were stark and apparent, and it was the nature of command to devise modalities to compensate for and even overcome them.

For mode (d), the attempt at control, there is little historical evidence of success—except when the predominance of force, or of technology, was such that neither error nor uncertainty could much affect the outcome. The many historical cases of failure include World War I as the most powerful, and tragic, example, with the British attack at the Somme its apotheosis.[14] Even in the Second World War, with its global scale and its gigantic operations, planning and coordination were contingent, and attempts at central control, when exercised, were usually disasters.[15]

At the end of the Second World War, things stood pretty much as they had since the turn of the century. The uncertainty-reducing properties of qualitative and quantitative advances in communications systems and information and data processing capabilities had been neatly canceled out by the increasing rapidity and complexity of combat operations. The greatest combat successes of the war had been managed at the "operational" level by great commanders such as von Manstein and Patton—as often as not "in spite of" attempts by theater and overall commanders to exercise control over them. The technological trends in communications and data processing started in that war were ultimately to revolutionize the role of command and control systems in the military. Ironically, such systems were to move far more quickly than the nature of battle, or the weapons themselves, leading, in Vietnam, to a whole new class of failures of over-control.[16]

It is, however, no longer possible to distinguish between command and control systems and the weapons themselves. The electronics revolution is no longer distinct from weapons and weapons system development. Indeed, the latter more and more depend intimately upon electronics and internal data processing for their very function. This is one aspect in which technologies of control have become intrinsic, creating a whole new class of weapons and systems that are expressions of rather than tools of information and data processing systems.[17]

Yet, there remains still the fundamental, and extrinsic problem of exercising command and control over these smart weapons systems. And here, the rapid advance in technology continues to escalate the problem of weapons and system control at least as fast as it has increased the capability of the central command structure. Technological change in the modern military, even in some of the less advanced countries, has increased both scale and geographical scope of operations far beyond the capabilities of previous forms of command-and-control systems to act in real time. In an era of supersonic aircraft armed with high-speed missiles, quick-reacting radar-directed gun and missile batteries, and tank battles that may be won or lost on the first shot, there is simply not the time for centralized command systems to exercise real-time control over battlefield events.[18]

Faced with the problem of operating such a critical large technical organization in situations where trial-and-error techniques are probably inapplicable because feedback and correction times are much longer than the time scale of the source events, commanders seem increasingly reluctant merely to advance and augment historical communications techniques, with their "management" orientation. Instead, they are adapting the new and powerful technologies of information and data processing to extend throughout their organizations a series of direct links for information and control, and placing at the center of the resulting web powerful, centralized command centers that are intended to exercise direct control from the top down to even the smallest of battlefield units.[19]

For these systems, as for all others, errors of various kinds will occur. However, as the technology in use grows in importance and scale, and comes to couple different units and different levels of command more tightly, the locus, nature, and scope of errors will change significantly.[20] Because of the increased coupling, misapprehensions and faulty performance are not so easily localized and repaired. And, to the extent that modalities of control come to dominate, the consequential deviations from modeled behavior can be far more profound. In some cases, the result may be a relatively complete collapse of the organization's performance, or of its ability to further pursue its primary objectives or carry out its primary mission.

To study this problem in large, to deal with strategic nuclear weapons, plans for the integrated AirLand Battle in Europe, or global movements of navies, is extraordinarily difficult, not only because of their multidimensional complexity, but because they have never really been tested (for which we are appropriately grateful). Thus, arguments about capabilities often turn into duels of competing ideology, with little evidence mustered on either side. Yet, there are a few, admittedly small-scale, cases where modern weapons systems and the approach to their use have actually been tested. The incident of 3 July 1988 *may* thus serve to illustrate in detail the type of comprehensive organizational failure that is likely to plague all attempts to exert "control" over large military technical systems.

The Scene Aboard the USS Vincennes

In the preface I characterized the situation in the CIC aboard the *Vincennes* as one of confusion and disorder. This may seem a harsh judgment, but it is borne out by the transcript of the Navy's official Investigation Report and, in retrospect, by the divergence between the memories and oral testimony of the officers present and the actual, recorded sequence of events.[22]

In the aftermath of the attack, and in its effort to justify, or at least forgive, the shooting down of an unarmed, civilian aircraft, any number of suggestions of Iranian culpability were advanced; that the aircraft had deliberately strayed off its course; that it heard, but would not reply to, the *Vincennes*'s challenges; even that it had a nefarious mission. Since the "black box" flight recorder on board Iran Air Flight 655 has been irrecoverably lost in the waters of the Persian Gulf, we shall never know exactly what her flight profile was, whether the crew ignored the American challenges or simply did not hear them, or whether they were aware of the fight going on beneath them or the impending attack.

We are, however, fortunate that as part of its advanced electronic suite, the *Vincennes* also had a sort of black box of its own. The SPY-1A, Command and Decision, and Weapons Control System computers were all equipped with magnetic tape equipment that tracked and recorded all of the signals received and processed by these key pieces of electronic equipment.[23] Because of this, we have been able not only to verify the timing and nature of all of her actions, but also to obtain a remarkable, perhaps unique, cross-check of the way that scenario preconceptions can distort both perceptions at the time of the event and memory afterwards, even when the supposedly causal stress has been removed.

The story told by the data tapes is straightforward. Iran Air Flight 655 took off from Bandar Abbas at 10:17 A.M. on the morning of 3 July on a heading of 210 (runway 21). Squawking Mode III, Code 6760 continuously, it kept on a more or less constant heading of 210, climbing steadily to its cruising altitude while gradually gaining speed. Data and testimony from the USS *Sides* corroborate the flight path and the Mode III IFF squawk. Indeed, the *Sides* was to identify the unknown aircraft as non-hostile and turn its attention elsewhere only seconds before the *Vincennes* launched its missiles.[24]

The story told by those inside the CIC aboard the *Vincennes* is quite different. From the first alerted contact, various personnel began to report a Mode II squawk, on a code (Code 1100) that was associated with Iranian F-14s. Although none of the data recorders reported any IFF response other than Mode III, Code 6760, those aboard the *Vincennes* continued to report Mode II, and to consistently mis-report the code of the Mode III signal. As the range closed, the *Vincennes* began to broadcast increasingly urgent warning messages to the unknown aircraft; at first, these were general challenges on both military (MAD) and international civil (IAD) distress nets. But as the notion that the aircraft was indeed an F-14 became fixed in the minds of the key operators, the challenges were made more specific—on the MAD net only, and addressed to an unidentified "Iranian F-14." A quick thumb-through of a listing of commercial flights missed the clear listing for Flight 655, although it was on course and nearly on time.

A warning of possible COMAIR (commercial aircraft) a minute or two later was acknowledged by the CO, but essentially ignored. At this point, the ship was still engaging the Iranian surface boats. Moreover, the ship was heeling sharply as the TAO (Tactical Action Officer) ordered 30 degrees of rudder at high speed to unmask the aft gun mount; loose books and equipment flew about the CIC. With the TAO concentrating on the surface battle and his attention divided, the CO depended on the Anti-Air Warfare

Commander (AAWC) to take care of the possible air threat. But the AAWC was new to his post, and generally regarded as inexperienced and a weak leader. *De facto* leadership fell upon the more junior Tactical Information Coordinator (TIC), who by that time was almost literally shouting about the immediacy and seriousness of the threat.

To give Capt. Rogers credit, he did allow the unknown aircraft to close to well within its possible missile firing range before asking for and receiving permission to intercept, and he did so only after repeating the challenge several more times. Only then, convinced that the threat to his ship was too serious to ignore, and under pressure to act quickly to avoid the earlier fate of the USS *Stark*, did he authorize the firing.

There are, then, three quite different narratives—that of the memory and perceptions of the officers and crew in the CIC, that recorded by the unforgiving instruments, and that described in report of the USS *Sides*, which was also monitoring the flight. But the *Sides*, after following the track for the same time, accepted it as a civil airliner, and turned its attention away *prior* to the *Vincennes*'s order to fire. In precisely the same circumstances, the high-technology command and control system "failed" to provide the means for a correct identification, while the low-technology one did not.

Was There an Error, and at What Level?

Was the CO justified in his perception of a real threat to his ship (which was the Navy's claim)? Were the Iranians reckless for flying over a fire fight in progress (which they may have been—*if* those at Bandar Abbas were aware of the exact position of the fight)? Or was the whole incident a regrettable, but unavoidable, accident of war (which is precisely what the resulting U.S. attitude was, in the Pentagon, in Congress, and in the press)?

The first question to be asked, then, is: Was an error made at all? The U.S. Navy was finally to claim that Capt. Rogers of the *Vincennes* acted correctly in appraising the threat, and ultimately awarded him a medal for his performance in the Gulf that day.[25] Others in the United States asserted that such blame as there was attached solely to Iran.[26] Iran, on the other hand, went so far as to claim that the United States had shot the Airbus down in full knowledge that it was a civil aircraft.

To this I respond that a *failure* did occur. The large technical military system operating in the Persian Gulf that day, of which the *Vincennes* was the central feature, was not waging total war, but rather a highly selective engagement in an arena known to be filled with civil traffic on air and sea. This very sophisticated piece of equipment had been placed in a situation for which it had never been designed precisely because it was thought to be most capable of making the kinds of quick and accurate judgments that would be necessary. It failed.

The second question to be asked is therefore: Given that the *Vincennes* is judged to have "failed" in that primary mission, was an "error" made, and if so, by whom? Linked to that line of inquiry is yet a third question: Having identified the primary error, was its source at the personal level (i.e., the performance of the CO or the TAO), the group level (i.e., collective failure in the CIC), or at broader and more systemic level (in which the actors, individually and collectively, were placed in a situation in which the probability that they could exercise proper judgment was greatly reduced)?

What links these two latter questions, and makes them most problematic, is the question of what constitutes an "error" at all. Properly speaking, an "error" is an individual act, whether of omission or commission, that takes place within a framework that provides sufficient interpretive context. When the root cause, however, is a misunderstanding of the framework in which the action was taken, the act may better be called a "mistake."[27] But if the decision framework and context as presented foster such misapprehension, the "mistake" lies at the operator level, the "error" at the organizational or systemic one.

The Investigation Report

Navy hearing boards such as the one convened to review the events of 3 July 1988 are unique in focusing on the CO, owing to the long Navy tradition that it is the CO, and only the CO, who is responsible for anything and everything that happens on or to his ship. As often as not, judgments are made on the simple basis of what the hearing officers would have done in the same situation, given the same information.

This Navy tradition worked against the possibility of a comprehensive, systematic investigation into the circumstances preceding the missile firing.[28] For the question should not have been whether the CO was justified in taking the actions he did given the situation and the information he had, but *how* the situation had developed so badly and *why* the information being provided was so skewed from reality. These matters were in fact addressed by the investigation, but by no means to the degree or depth that would have been required to develop an adequate set of answers.

The Investigation Board was convened by Rear Admiral William M. Fogarty at Bahrain beginning on 6 July, while the events were still fresh in the minds of the participants. Formal hearings began a week later, and the entire procedure was completed and the report delivered to the Navy on 28 July.[29] Even in the cleansed form provided to the public, the report is rich in personal and technical detail. Perhaps the most striking feature is the degree to which the recollections of the participants as to the nature and assessment of the presumptive threat differ, and the variance between what was reported by the SPY-1A computers and what its human interpreters were reporting.

The record shows that the decision to fire was taken more or less calmly and deliberately on the basis of personal advice passed from junior officers to the senior AAWC (Anti-Air Warfare Commander), and from the AAWC to the CO—in the face of a stream of contrary evidence from the electronics aboard. Faced with the problem of reconciling the manifest mistakes made in interpretation of technical evidence, the Hearing Board concluded that "stress, task-fixation, and unconscious distortion of data may have played a major role in this incident." The report then went on to attribute the distortion to the relatively junior TIC and Identification Supervisor (IDS), who became convinced that the track of Iran Air 655 was an F-14 after an IDS report of a momentary

Mode II squawk. The Fogarty report states:

> After this report of the Mode II, TIC appears to have distorted data flow in an unconscious attempt to make available evidence fit a preconceived scenario ("scenario fulfillment").

This fulfillment included continuing to read the Iran Air flight as descending toward the *Vincennes*, even though the information being presented by the electronic suite was contradictory.

Because it was perceived that the AAWC was a weak and inexperienced officer, the TIC took effective leadership, and his reports were accepted by all. His reports were also heard by all, since he kept reporting the closing of the "hostile F-14" repeatedly over an open communication circuit, and was practically shouting at the time the missiles were launched. In such circumstances, it may indeed be remarkable that the CO, deprived of any direct source of information or data, was able to delay his decision so long.

Whether or not he would have been justified in shooting if the evidence were simply ambiguous or the uncertainties known is not our central concern here. Rather, the focus of our analysis is the manifest failure of the decision-making system on the *Vincennes* to interpret the evidence correctly. The system "failed" in that a false hypothesis was constructed and maintained, serving as the basis for all subsequent actions. This would have been as serious a matter even if Capt. Rogers had decided *not* to fire, although it is most likely that we would never have learned about it.

Among the conclusions of the Fogarty Report is a general recommendation that

> Since it appears that combat induced stress on personnel may have played a significant role in this incident, it is recommended the CNO [Chief of Naval Operations] direct further study be undertaken into the stress factors impacting on personnel in modern warships with highly sophisticated command, control, communications, and intelligence systems, such as AEGIS. This study should also address the possibility of establishing a psychological profile for personnel who must function in this environment.

Stripped of its deliberately restrained prose, this is a quite remarkable admission that the very sophistication of the C3I system may in itself have been a contributory factor in the incident. This is the only place in the official report where there is any acknowledgment that it might be wise to differentiate between questions of personnel performance and training at the individual and/or group/ship level and more systemic factors.

In summarizing the Fogarty Report in his endorsement, the Commander in Chief, U.S. Central Command, pointed again to the questions of stress and time compression, and the intervention and near-hysteria of the TIC. However, he also criticized directly the performance of the AAWC for failing to confirm the reports on his own, or to check the visual display on the console directly in front of him (which would have shown the track to be ascending and not descending). It would appear that he, too, concurred with the scenario that was becoming fixed in the CIC, and trusted his junior officers for information instead of checking the system independently. This endorsement carried a recommendation for the issuance of a non-punitive letter of censure to the AAWC, a recommendation that was overturned by Secretary of Defense Carlucci prior to the press conference of 19 August. At that conference, Admiral Crowe, Chairman of the Joint Chiefs of Staff, admitted that "mistakes were made on board the *Vincennes* that day," but finally reported that, in the opinion of the Navy, there was no *culpable* conduct displayed. The event was a very regrettable accident, but basically a by-product of the Iran-Iraq war.[31]

It should also be noted that the entire review process was treated as one investigating "culpability," as if personal malfeasance or neglect was the only issue. Having absolved the individuals of blame, the Navy, and, by and large, the press and Congress, then moved rapidly to the highest possible level of political analysis: if there was no personal malfeasance, and the Aegis system had worked perfectly, the result could only be attributed to the misfortunes of war.

The APA Panel

On 6 October 1988, a panel of five psychologists chosen through the American Psychological Association (APA) testified at a hearing of the House Armed Services Committee, accompanied by a great deal of media coverage.[32] Although a large part of their testimony was directed toward the question of decision-making under the stress of the circumstances, and at the failure of the Office of Naval Research and other federal agencies to adequately support research into decision-making under stress, they also raised at the hearings the more general question of the techno-systemic environment within which the decisions were being made.

As the Fogarty report had done in its summary, the expert panel pointed out that in an era of increasing technical "complexification," it will no longer do to continue to point only to "operator error" as the source of malfunctions and disasters.[33] Rather, what happened aboard the *Vincennes* on July 3 could be seen as part of what one psychologist characterized as the "glass cockpit" syndrome, which in itself is a property of large, highly sophisticated technical systems operating under conditions of high stress, high consequence, and high visibility.[34] Unfortunately, neither the Armed Services Committee, the Navy, nor the Department of Defense seems to have had the least inclination to follow up this line of inquiry.

Facts and Factoids: Of Relevant and Irrelevant Argument

Before engaging in a further analysis of the performance of the *Vincennes* as a weapons system, it would be well to deal first with some other assumptions and beliefs about the incident that serve only to draw attention away from important matters.

Of these, the most public, and striking, is the attempt to attach some or all of the blame to the Iranians. Of course, we have no way of knowing what the pilot or crew of Iran Air Flight 655 did or did not know about the fire fight going on below their assigned corridor, or why they (unluckily?) chose to fly to the north of the centerline instead of the south. Nor, given relations between the United States and Iran, do we have any way of directly interviewing the staff and control tower at Bandar Abbas, to see what they knew. It is true that *if* Iran Air Flight 655 had been directed away from the fight, or *if* it had left on time, or perhaps fifteen minutes later, or *if* it had been monitoring the MAD or IAD channels and responded, or, more speculatively, *if* the *Vincennes* had not fouled its mount and gone into a high-speed turn, the tragedy would probably have been avoided.

Much attention was also paid to whether the very expensive and highly sophisticated Aegis radar system did work, technically speaking. In fact, it did work—in the sense of identifying and correctly tracking Iran Air Flight 655 almost from the moment of its takeoff. Given its technical limitations, it did not, and could not, identify the type of aircraft, or even its size. And the rest of the electronics suite did correctly identify the IFF signal as Mode III Code 6760, and *only* Mode III Code 6760, throughout the brief flight.

But as subsequent events demonstrated, the system clearly was open to misinterpretation by its operators, and failed to provide strong enough or clear enough counter-evidence to overcome their mindset once it was fixed. Thus the question of the functioning of the *Vincennes* electronic hardware (and software) *as* hardware is simply another irrelevancy to this particular series of events. More important by far is that the *Vincennes* failed as an operating organization, however perfectly its equipment was operating.[35]

A third major point raised was the question of "stress." It was emphasized during the official investigation and in the reports of the five APA psychologists that stress played a major role in the tragedy. Indeed, the whole idea of "scenario fixation," in which the Iran Air flight became embedded in the collective consciousness of those in the CIC as a hostile Iranian F-14, is derived from and attributed to the degree of stress then present. The Official Airline Guide was aboard the ship, and Iran Air Flight 655 was listed, yet the entry was not found. The *Vincennes* claimed to have detected a Mode II signal identifying an Iranian F-14 (which in any case is not, to anyone's knowledge, fitted to carry anti-ship missiles), and thereafter acted as if the identification were firm, to the extent of ignoring or misreading evidence incompatible with that scenario. Was this the significant error?

The testimony given before the Investigation Board was almost unanimous in stating that stress in CIC was very high at that point, although there is no way retroactively to determine just how high it was, or whether the sense of near-panic so clearly evinced by the TIC was spreading. But again, what are the expectations of a combat system such as the *Vincennes*, or of the CIC as a war-fighting center? That stress would be low? That battle conditions would be other than confusing? That the ship could be attacked on the surface, or from the air but not both simultaneously—not to mention possible sub-surface attacks in other circumstances? If these are or were the assumptions under which the Aegis cruisers were designed, than the *Vincennes* is useless as a fighting ship.

These are not major issues but distractions, not so much facts as "factoids." They have the appearance of important facts but lack analytic substance. It is true that it took a series of circumstances to bring the Iran Air flight into range at a critical time. But this matters only if the major point is to find out why the risk, the potential for tragedy came about. Since our objective here is primarily to find out why the risk turned to reality, they are not relevant to our purpose. And neither, to a great extent, is the matter of stress.

Stress

The two primary factors identified in the hearings, the reports of the eminent psychologists, and most thoughtful journalistic analyses were the misreading of the commercial aircraft as descending when it was in fact ascending, and the persistent failure to correctly read out the IFF data being presented by the electronics suite in the CIC. Both of these are then attributed to the rather considerable stress at a time when the CIC was otherwise somewhat disrupted from the preceding abrupt maneuvers.

Subsumed under the category of stress as well is the institutional memory of the attack on the USS *Stark* the previous year. Within the Navy, the presumptive "lessons learned" from the *Stark* incident were sufficiently strong, and sufficiently unambiguous, that "the USS *Stark*" is simply listed in the Investigative Report as one of the contributing causal factors without any further expansive discussion.

Stress was clearly a contributing factor, but its presence is in no way explanatory, by no means exculpatory. Stress is, or should be, a central assumption of battle, along with confusion, uncertainty, and risk. To design our systems otherwise would be sheerest folly. Based upon our long naval history, there is simply no evidence that the *Vincennes* was under a degree of stress greater than one would expect for a ship in combat, in strange waters. If the ship as a system is incapable of operating correctly under such circumstances, she is a failure as a weapons system, however well her machinery, electronics, and missiles perform.

The *Stark* failed to identify the Iraqi aircraft as potentially hostile, and failed to arm and unmask its missile defense systems properly. But in what way did that contribute to the sequence of events aboard the *Vincennes*? The *Stark* was not prepared for hostile attack, not suspicious enough of a potentially threatening flight profile, and not ready for combat soon enough in the chain of events. The *Vincennes*, however, was prepared, was suspicious, and was ready. Clearly it was not the specifics of the *Stark* incident that laid combat stress on the officers and crew of the *Vincennes*, but rather the memory of the consequences of any serious failure at all.

It is in this context that the matter of scenario fixation returns as a function of the shift from technologies that assist battle management to those that attempt to control the battle. Aboard the *Sides*, the officers and crew were free to experiment with scenarios, to use their data processing and information-gathering equipment to assist with alternative descriptions of reality until they found one that seemed best to fit the circumstances. This is management, in its classic form, with technological systems serving their proper roles as inputs and auxiliaries.

Aboard the *Vincennes*, the role and nature of the technology was quite different. Whereas the *Sides* remained an individual ship, trying to fit its actions into a larger pattern in modes reminiscent of military commanders past and present, the *Vincennes* saw itself as part of larger and more integrated systems of C_3I, as a key element in a large technical system extending, in this case, over the entire Gulf. Moreover, the Aegis system and its auxiliary equipment had programmed into it the kinds of scenarios for which the system was designed. It is then not surprising that a skilled officer cadre and a highly trained crew, extensively trained and practiced in exercises and on simulators to play the "games" for which the *Vincennes* was designed, should have fallen back, under stress and severe time pressure, into one of the many pre-programmed behavioral patterns.[38]

Toward a Typology of Errors

Organizations operating in an environment of high technical or operational risk ultimately depend for their performance upon human operators. The organizational design is geared to producing an individual who will always make the "right" decision in a critical situation.[39] Therefore, if an accident does occur, and was not beyond the operator's perceived range of control, the organization can reach one of two conclusions: (a) the organization failed in its design, placing the right operator in the wrong situation (or the wrong operator in the right situation); or (b) the operator failed as an individual, despite organizational design and training. It is clearly more desirable for the organization to believe in (b), the fallibility of the individual, than in (a), its own imperfections.[40]

The classic term for (b) is "pilot error," which is in itself sufficiently broad to warrant further explication. In the early days of passenger flying, it was assumed that there was a fair portion of risk involved, and that the causal factors were not only uncertain and difficult to anticipate, but perhaps unknowable in advance of actual circumstances. The job of the pilot was not just to provide technical skill as a machinery operator, or complex guidance by means of skills and techniques not generally known, but also to cope with the unanticipated, and threatening, as it occurred. Thus, when accidents did occur, the blame was almost always placed on pilot error—except when the physical or technical malfunction was so blatant, and so unambiguous, that no action by the pilot could possibly have saved the situation.

In the case of fairly large, fairly complex organizations operating increasingly sophisticated technologies, the category of individual error as described above has to be further expanded to allow for the fact that some operations must be performed by a group of people acting in close concert. Thus, "pilot error" as a concept can be extended to "group error," where the group in question can be the team in the cockpit of an advanced airliner, the personnel on the bridge of a large ship, or the team in the CIC on an Aegis cruiser.[41]

Even when so expanded, "pilot error," remains a portmanteau, subsuming two general categories of presumptive malfeasance: (a) failure to operate the equipment properly, or skillfully, or to follow various procedures and rules established to guarantee operational safety; and (b) failure to rescue operations from an unanticipated or abnormal situation that was nevertheless within the presumed skill and capacity of the expert operator to rectify or remedy.

Following the terminology of Martin Landau, we refer to the first of these as a "Type I" error—overlooking, ignoring, or misunderstanding the information presented even when it occurs within the envelope of the predicted or anticipated flow of events.[42] It should be noted that this class of error also includes "anticipatable" equipment malfunction or stress over a wide range, such as engine failures, electronic malfunctions, or high-stress operational periods such as take-offs or surface combat. Included in this class, therefore, is the failure to rectify or save a variety of situations that are presumed to lie within the range of an expert operator to correct or act upon because of his or her skill and expertise. The CO and officers of the USS *Stark*, for example, committed a Type I error in failing to put the ship on proper alert when an Iraqi aircraft was sighted.

The second we will call "Type II" errors: accepting as true, accurate, or significant information that is misleading, incorrect, or irrelevant, or, by extension, projecting into a situation "external" beliefs or assumptions about the nature of the situation or the state of the system. One example might be the failure to set the flaps or landing gear into proper position during take-off, or to notice that de-iced wings have re-iced; persistent misidentification of a potentially hostile aircraft heading toward a naval unit in a combat zone would be another. The actions of the *Vincennes* might well be so classified.

However, these historical classifications are increasingly misleading for large technical systems, particularly those involving some degree of actual (physical) risk. The systematic progression and "complexification" of advanced technology has resulted in incidents that were once classified as "Type I" moving progressively toward "Type II." Redundancies in equipment, presumptive higher reliability, and considerable sums spent on system design and large-scale integration result in the anticipation that such failures will be increasingly rare. One result is that pilots (operators) will acquire very little skill or experience with the acceptance and interpretation of signals of potential error. Lacking experience with complex systems whose informational modalities require formal interpretive training, the operators will find it very difficult not only to filter the information they are getting, but to discriminate between their intuitive or prejudicial judgments and those that the equipment is supposed to provide.[43]

The very complexity of the equipment will make it more difficult to ascertain what is going awry, why, and, at times, whether something is going badly or incorrectly at all. Where a pilot could once try to make a determination as to why his radio was not working, and what he could do about it, there is no longer any airborne capability for repair, let alone for "string and sealing wax" patchwork remedies. A malfunctioning compass was more easily noted as in error than is a triply redundant inertial navigation system. As the equipment gets more complex, and presumably more reliable, the operator's job is also progressively redefined from that of error detector and manager of continually suspect equipment to that of an "administrator" of integrated systems whose internal functions are remote from his or her experience.[44]

But both of these classifications still omit one major consequence of advancing technological sophistication and operational complexity. As the equipment becomes ever more elaborate, and the operators move increasingly from hands-on to representational modes of analysis, there is always the danger that the representation presented to the operator is in itself incorrect and/or misleading. To the historical classes of errors of omission and commission as described above, we must therefore add a third class, referred to by Karl Weick as "errors of rendition."[45]

For large technical systems, the historical tendency may therefore be described in terms of motion: (a) from individual error to collective error at the team, group, or "system" level; (b) from Type I errors, which are essentially the responsibility of the operator, to Type II errors, whose causal roots increasingly lie at the level of equipment design and operational administration; or (c) toward a situation where representational errors arising from system design and modes of integration and administration may make the entire process of hypothesis formation sufficiently suspect that failures will occur even when the operators have no reason to suspect that *either* of the two classic types of error is being made.[46]

Of course, more traditional forms of error will still occur. The pilot/operator may simply fail to follow rules, or to operate the equipment safely or correctly. The task may be such that exaggerated assumptions were made as to operators' ability to take action to save an unanticipated situation. And small-group errors may in many circumstances also be treated fairly accurately as extensions and generalizations of individual behavior, with an expanded understanding of the role of the group in forming and accepting false hypotheses and interpretations.

But there is also a growing class of incidents for which the assertion that the operators "could have" or "should have" been able to properly detect, interpret, and remedy an incipient error carries with it the embedded assumption that the organization has provided a correct representation, sufficient training, and experience to judge whether or not the equipment is reliable, or correct under specified circumstances, and a decision-making framework in which the operator can separate judgment for the immediate situation from the role as representative of the larger organization. Thus, operators may well be placed in positions where there is no reliable post hoc way to evaluate whether they *could* have acted correctly within the parameters of the specific situation with the technology, equipment, and decision and control frameworks they were given.[48]

Technology and the Span of Command and Control

The *Vincennes* incident has provided us with a potentially valuable case, from which we may be able to draw some generalizations and conclusions about the future of long-range, highly automated, control-oriented, high-technology weapons systems. It is this potential that makes it so regrettable that the Navy's investigation so systematically sought to identify and localize specific, non-systemic factors upon which the "blame" (to the extent that it could be characterized as blame at all) could be placed.

In the days of wooden ships, command was relatively straightforward. Threats were clear and simple, developed at a reasonable pace and usually within sighting range, and were confined to the plane of the surface. A modern fighting ship fights in a three-dimensional environment that extends to hundreds of miles, facing threats that may develop with great rapidity, and with a demand to interpret and integrate many disparate sources of data and information. The "threat environment" so defined is too complex to be the responsibility of the ship's CO, who must incorporate the threat into an even larger picture concerning the well-being of his ship, its relation to others, and communication with higher command. But combat decision-making is not a function that can be decentralized when threats and decisions develop rapidly. Aboard modern U.S. ships, the task of actually "fighting the ship" is assigned to the TAO.

Over the past few years, I have had the privilege of being part of a research group that was allowed aboard several U.S. aircraft carriers, as part of a project on the organization and management of high-risk technological operations.[49] During that time, we have observed Combat Decision Center (CDC, formerly known, as on Aegis cruisers, as the CIC) operations at sea. Although we did not, and could not have, been in the CDC during real, or even potential hostilities, we have been present as observers during extensive and realistic training, and during combat exercises in Pacific waters that serve the dual role of final training and evaluation before the ship is qualified for combat and deployed.

Owing to the complexity of the combat environment for a carrier, which may at different times act almost alone, in concert with only a few escorts, or as part of a larger Battle Group, the tactical functions are all available and drilled aboard the ship itself. Thus, there are four main combat activities that need coordination and control: anti-air warfare (AAW), anti-surface warfare (ASuW), anti-submarine warfare (ASW), and offensive coordination (Strike). Of the four, the first two—anti-air and anti-surface warfare (primarily defensive)—are closest in specification to those aboard the *Vincennes*, and will be used here for purposes of comparison.[50]

As on the *Vincennes*, AAW and ASuW coordination each center on a single person—the TAO—charged with performing the integrative tasks of assimilating different information from a variety of sources, assessing the current state of threat, and placing it into spatial and temporal context. In the case of the aircraft carriers we have observed, there is some "senior oversight" of the TAO, who is usually a relatively junior officer. Behind him in the CDC are the ship's Combat Officer, possibly the Admiral's Combat Officer (if the Admiral is aboard), and perhaps one or two other senior officers from ship and/or Admiral's staff.

But the primary job of these senior personnel is to provide the line of authority to the TAO, who cannot, on his own, launch or divert aircraft or release weapons systems. They will also be coordinating the different "warfare" officers, each of whom is acting within his own domain (e.g., surface, subsurface, air). Sometimes they will all be in direct reach; at other times, they may be in remote locations, so that the senior staff will also serve as channels of communication as needed. In addition, they try when able to give general guidance to the TAO, and to serve as conduits for other kinds of information (e.g., intelligence) not displayed on the consoles and status boards in front of him. They do not have the sense of the four-dimensional threat environment in which the ship is operating—that integrative task falls solely on the TAO, and it is a formidable one.

Past and present TAOs have characterized their sense of having proper command and integration of the information flows as "having the bubble." When you've got the bubble, all of the charts, the radar displays, the information from console operators, and the inputs from others and from the senior staff fall into place as parts of a large, coherent picture. Given the large amount of information, and the critical nature of the task, keeping the bubble is a considerable strain. On many ships, TAO shifts are held to no more than two hours. When for one reason or other the TAO loses the sense of coherence, or cannot integrate the data, he announces loudly to all that he has "lost the bubble" and needs either replacement or time to rebuild it. Losing the bubble is a serious, and ever-present, threat, and has become incorporated into the general conversation of TAOs as representing a state of incomprehension or misunderstanding even in an ambience of good information.

This notional "bubble" is one of the key elements in obtaining high and reliable systemic performance in organizational settings that involve considerable risk, large and basically uncontrollable uncertainties, and a high degree of organizational as well as technical complexity, operating in real time situations where decisions with irrevocable consequences must be made quickly, in real time, on the spot.[52] At the cognitive level, the conception of the "bubble" makes it possible for Navy TAOs, Air Traffic Controllers, and (in a singular and somewhat different context) pilots in melee combat to integrate a system of vast operational, technical, and cognitive complexity into a single, mental, representative spatio-temporal picture or map that organizes and orders the flow of information and allows decisions to be made on the basis of overall, systemic situations and requirements.

As might be expected from the description, bubbles cannot be acquired quickly or simply. Aboard the ships we have observed, TAO shifts overlap by up to an hour to make sure that the "bubble" is transferred smoothly to the succeeding TAO without a potentially dangerous break in routine or perception. Bubbles are primarily human decision systems.

Moreover, the proceedings of the Investigative Board show little sign of any primary role for "bubble" formation aboard Aegis cruisers. Instead, the role of the multiple technical systems, which aboard a carrier serve primarily to feed information *to* the TAO, seems to have been reversed. Aboard the *Vincennes*, the Aegis-directed, computer-operated anti-aircraft missile system is the primary system, with the CO primarily involved in deciding whether to activate its firing mode or not. In this situation, the TAO has become an input, not an integrator, and the CO an authority rather than an expert.

Instead of a closed decision-making system, based on the historical notion of the integrity and isolation of a ship at sea, the CIC on modern surface ships is increasingly a part of a larger web of information, control, and authority; instead of working with technologies whose characteristics are deeply and intimately familiar to those aboard, decision-makers must increasingly rely on abstract displays from systems whose inner workings are opaque to them. The locus of authority and scale of consequence have rapidly expanded, while the responsibility remains at the operational level.

High-Technology Military Systems: The Two-Edged Sword

Taken in context, the events surrounding Iran Air Flight 655 are illustrative of a series of disquieting trends within large-scale military command and control systems. The *Sides*, with less sophisticated equipment, a less comprehensive command and control system, and subject to larger direct risks if attacked (owing to its less sophisticated defensive systems), made a correct evaluation. But it was also a less valuable, less attractive target, under less pressure to act as part of the larger system of air defense in the Gulf. The *Vincennes*, despite its awesome capabilities, felt more threatened, more vulnerable, and under more pressure. Indeed, one might even say that the degree of pressure, the fear of threat, and the perceptions of vulnerability were exacerbated by the awareness of her CO of the *Vincennes*'s cost, her value and irreplaceability as the core of air defense, and the expectations of those who bought her, built her, and sent her into the Gulf.

Moreover, the nature of her equipment, and the training to use it, put the crew very much in the mode of "control" rather than management. Instead of being under constant pressure to interpret data and adjust systems to meet changing realities, the *Vincennes* was prepared and trained to activate an elaborate system of control in a situation where the only uncertainties would be those deliberately created by potential enemies trying to confuse their information and data collection system. Furthermore, the manifest confusion aboard the *Vincennes* is absolutely no excuse for her poor performance. Military systems that cannot function in the midst of confusion, or while under multi-threat attack, are not only totally useless, they pose more dangers to their users than to their opponents.

More detailed attention should have been paid to the false cognitive map identified in the hearings that was so casually identified there and in the press as a form of "human error." When dealing with large technical systems, error comes in many forms having to do with judgment and cognition, not all of which have their origins at the operational level. A situation may not be identified correctly, or in time, or not properly evaluated. A responsive action may be taken incorrectly, or be inappropriate to the circumstances, or be tagged as wrong in retrospect even when it was probably the best judgment that could have been made at that time under those circumstances.[53] But the operators may also be placed in a situation where they have insufficient time to perform or evaluate the identification, insufficient experience to act on the basis of trained intuition, and a decision milieu that seriously preconditions or hampers their responses.[54]

Managers and commanders alike dread uncertainty and the inherent unpredictability of human actors—especially when situations are complex, the information flow high, and the time to act short. Under such circumstances, the temptation to try to control processes to bound outcomes is great. Attempts at intervention through such mechanisms as global C_3I systems, satellite data links, and elaborate, more or less automatic, pre-programmed, computerized control systems are therefore not surprising. Automatic systems can process data much faster and, it is promised, interpret it faster, more efficiently, and with fewer errors. What is generally overlooked is that such systems reify the models of the designers, imposing a modality of control as if a exhaustive, predictive knowledge base were in place.[55]

Fortunately, those who design and build military systems still acknowledge that the final decision, the decision to actually shoot, is far too important to leave to the machines, at least in times of peace. (There is, however, a continuing concern about delegation in time of actual war.) We are assured that there is still a human being who sits above, or below, or next to the automatic system and makes the ultimate decision. But on what evidence, and on what grounds will that decision be made in the future? Based on the evidence presented in the case of Iran Air Flight 655 (and on corollary evidence from other military incidents), the elaborate information and control system built into the *Vincennes* laid the very trap she fell into. It was a complete system, built with a number of assumptions that ended up creating in the minds of those in command a scenario compatible with the range of events the system was designed to respond to. In other words, under stress, and pressed for time, the CO and the TAO were not about to ask whether the circumstance they were in was one appropriate to the elaborate control system in which they were embedded, and for which they were responsible. Instead, they created and acted out a scenario compatible with the threat the system was designed to combat. A tragic error on their part, but not entirely an operational error.

Nor did the sophistication of their information and data processing system provide any real assistance. One of the crucial things that happens in a "bubble" is the constant building and reformation of scenarios of what is going on. Without a scenario—a script, if you will—the incoming information is too dense, and comes too fast, to be interpreted. Given the rate at which information was coming in, the CO and the CIC staff had no time to devise scenarios. Nor, given the increasing focus on systems of "control," were they trained in quick, heuristic scenario-building. Therefore, they reacted directly to the threat as a threat, within the script they were already playing.

As we move to more expensive, more capable, weapons systems, the demand for systemic control to avoid error will surely increase. Until recently, this was true primarily of nuclear weapons systems, where the cost of a mistaken launch, or release of weapons, was so clear, and so dramatic, that almost literally heroic measures have been taken to prevent it.[55] Less well-known, but equally important, have been the continuing concerns over the ability to fire the system when necessary, in an atmosphere steeped with the precautions against inadvertent or incorrect action.

Heretofore, the distinction between nuclear and non-nuclear weapons was so great that the notion of systematically evaluating the potential for both kinds of error was rarely, if ever, extended beyond the traditional point of the training and testing of individuals, particularly those in command. But the advent of other "conventional" systems as sophisticated, but without the terrible risk that accompanies the nuclear systems, implies that we will have to extend what we have learned to date. Moreover, many of these new conventional systems—such as SDI technologies and, as it turns out, Aegis and similar air defense systems—may well be geared for a quicker response time. Most will also be lacking the multiple levels of restraint that both available time and tremendous risk have imposed on strategic nuclear weapons.

Thus, the incident of Iran Air Flight 655, in context, serves as a sharp example of the perils of increasingly sophisticated, increasingly centralized, and increasingly expensive military C_3I systems. Whether they be bunkers under Cheyenne Mountain, AWACs flying over Western Europe, or forward command posts on either side of the inter-German border, these systems grow larger, more rigid, and more saturated with information and responsibility with every passing year. We are constantly reassured that the systems are workable, and only need more careful training of personnel to become error-free.

Even the *Vincennes* case is discounted, on the grounds that the Aegis system worked correctly. This is perhaps the most misleading statement of all, for without the Aegis system it is possible that the events described above would never have occurred. The Navy's complacency about the success of its hardware in the face of the failure of the *Vincennes* is analogous to saying that the operation was successful but the patient died. If the large social-organizational-technical system fails to produce the right outcome, then it has failed *as a system*. Whether the proximate cause was human action or malfunction of the equipment is a matter of detail; substantive detail, to be sure, but detail nonetheless. It does not absolve us of the need to inquire further to discern whether the inability of the operators to avoid a mistake or recover a deteriorating situation occurred because of the structure and context of the large technical system in which their decision was embedded.

NOTES

1. Committee on Armed Services 1987. For an excellent concise description, see Vlahos 1988.

2. *Ticonderoga*-class Aegis cruisers are essentially built on long, narrow destroyer hulls for speed and maneuverability, and roll considerably in high-speed turns.

3. Most of the detail in this section is obtained directly from U.S. Department of Defense 1988 (hereafter, the Fogarty Report).

4. Notably excepting Chris Demchak in her remarkable study of the introduction of the M-1 tank as a socially as well as a technically complex "system" into the U.S. Army.

5. For many of the more sophisticated social historians of the military, the historical problems of command have always entailed the discussion and analysis of decision-making under the most profound uncertainty. Yet, the costs and risks of warfare and battle are such that the uncertainty, however inevitable, however unavoidable, is never tolerable, and rarely accepted. Thus, Martin van Creveld's magisterial history (van Creveld 1985) has centered almost entirely on how certain famous (or infamous) commanders did or did not attempt to take "control" over events in their "quest for certainty."

6. Rasmussen and Batstone 1988.

7. Some examples include van Creveld 1985 and 1989. Among the many authors operating in more traditional modes, the works of John Keegan stand out as best portraying men in combat in the rich complexity of their organizational and unit setting. Of the behavioral and sociological works, the most noteworthy were originally funded as studies of the Army in Vietnam. See for

example Sarkesian 1980.

8. See O'Connell 1989. Several authors have made this point within the more restricted realm of computer technology; see, for example, Bellin and Chapman 1987.

9. See March and Weissinger-Baylon 1986.

10. Compare, for example, Beninger 1986 with the definitions used by military analysts or by Landau and Stout (1979).

11. Landau and Stout 1979.

12. van Creveld 1985; see also Keegan 1987. The definition of complexity as measured by the number of units, their differentiation, and the interdependence between them derives from La Porte 1975. Scale is an arbitrary term having to do with both geographic and conceptual scope.

13. Excellent discussions of Moltke and the Prussian General Staff can be found in van Creveld 1985 and Keegan 1987.

14. van Creveld 1985, pp. 155ff., is particularly good on the rigid, inflexible timetable planning of the artillery barrage that left the British troops at the Somme uncovered when their progress was "delayed."

15. The increasing attempts by Adolph Hitler to manage the front directly as World War II dragged on are the most famous, but Allied commanders were susceptible to the same temptation. See, for example, the several World War II histories of Liddell Hart, and the memoirs of, for instance, von Manstein and Rommel.

16. van Creveld 1985, pp. 232ff. Gibson 1988 may be the single best book on how American managerial and technocratic assumptions led to a systemic defeat. For analysis of a more recent episode of overcontrol see Wilson 1986.

17. Bellin and Chapman 1987; Din 1987; Deitchman 1983.

18. See, for example, Lanir et al. 1988.

19. Canan 1988.

20. The term tight-coupling is most precisely delineated in this context in Perrow 1984. The terminology is largely derived from Thompson 1967. See also Weick 1976.

21. I hesitate here to use the term "systemic" failure," since to some that implies a failure of the Navy as a system, rather than the ship. Moreover, the ship as a system can only be understood as embedded in the larger context of the Persian Gulf command, which in turn is embedded in the still larger context of Fleet and Area commands, and so on. Other relatively complex large technical systems may also share this property of systemic "nesting," which complicates description more than it does analysis.

22. This was much clearer in the perusal of the Fogarty Report (see note 3) than it was in subsequent news stories.

23. The Aegis Display System data could not be extracted, precluding any positive confirmation of actions taken at the CO and TAO consoles.

24. To quote the Fogarty Report directly: "The data from USS *Vincennes* tapes, information from USS *Sides* and reliable intelligence information, corroborate the fact that [Iran Air Flight 655] was on a normal commercial air flight plan profile, in the assigned airway, squawking Mode III 6760, on a continuous ascent in altitude from takeoff at Bandar Abbas to shoot down."

25. *Washington Post*, 30 April 1990. The Legion of Merit, the U.S. armed forces' second highest award, was presented to Capt. Rogers and Lt. Cmdr. Lustig (the weapons officer) for "meritorious conduct" and "heroic achievement" on 3 July 1988. The citations did not mention the downing of the Iran Air flight at all.

26. Admiral Crowe was reported to have concluded that airport officials at Bandar Abbas "should not have allowed the flight to take off while a fire fight was going on some miles away" (*New York Times*, 20 August 1988). Then-Vice-President George Bush was quoted as saying: "I will never apologize for the United States—I don't care what the facts are." For these and similar quotes, see, for example, George Wilson, "The Risks of Shooting First" (*Washington Post*, 28 January 1989). To be fair to the Navy, their conclusion that Capt. Rogers was justified in firing in presumptive self-defense was widely shared on Capitol Hill. See, for example, *Congressional Quarterly*, 9 July 1988, 1905ff.

27. As Karl Weick has put it, "An 'error' occurs when a person strays from a guide or prescribed course of action, through inadvertence and is blameworthy, whereas a 'mistake' occurs when there is a misconception, misidentification, or misunderstanding" (Weick 1988).

28. This arises from the usually praiseworthy desire to expedite inquiry while evidence is fresh and untainted. The problem lies not with the speed of the first inquiry, but with the closing of the book on the entire incident once this hurried and summary court, held in the combat area, far from other sources of information and analysis, is concluded.

29. The version of the Fogarty Report released to the press on 19 August was cleansed of classified material and of potentially damaging references to specific individuals.

30. Fogarty Report, p. 45.

31. Transcript of a news briefing by Secretary of Defense Frank C. Carlucci and Chairman of the Joint Chiefs of Staff Admiral William J. Crowe, Jr. at the Pentagon, Friday 19 August 1988, 11:00 A.M.

32. Baruch Fischhoff, Robert Helmreich, Richard Nisbett, Richard Pew, and Paul Slovic each presented testimony "on behalf of the American Psychological Association."

33. Testimony of Baruch Fischoff before the Committee on Armed Services, U.S. House of Representatives on behalf of the American Psychological Association, 6 October 1988.

34. Testimony of Robert L. Helmreich before the Committee on Armed Services, U.S. House of Representatives on behalf of the American Psychological Association, 6 October 1988. See also Squires 1988.

35. See Perrow 1986. This crucial distinction is rarely made in the policy debate over the military and its equipment, or even for analogous systems such as nuclear power plants or chemical refineries.

36. Alas, the Aegis cruisers may well have been so designed. The CIC is located high up in the superstructure, behind the Aegis radar panels rather than in the main hull. Aegis cruisers, with their light displacement and narrow beam, roll easily, and the CIC location multiplies the effects of the ship's motion on those fighting it. Their primary mission is that of long-range air defense of an aircraft carrier Battle Group in blue-water operations. Thus, they represent yet another move in naval warfare away from the close grappling of ancient days through cannon, rifled guns, and aircraft to remote combat fought at extreme ranges by electronics.

37. Byron 1987.

38. See U. S. General Accounting Office 1988 for a critique of the Pentagon's failure to test systems under realistic stress. See also Byron 1987.

39. Rochlin, La Porte, and Roberts 1987.

40. Rasmussen 1989; La Porte 1982.

41. Roberts (forthcoming) provides an excellent example drawing on the U.S. Navy. Also see Rochlin, La Porte, and Roberts 1987.

42. Landau and Stout 1979. This article defines two basic types of organizational error—"Type I" error (consisting of rejecting as false a hypothesis that is true), and "Type II" error (consisting of accepting as true a hypothesis that is false). This in turn is based on the fundamental conception of an organization as a knowledge-seeking entity. Strictly speaking, however, the terminology of Type I and Type II should be used primarily to discuss organizational rather than individual performance.

43. Rasmussen 1986.

44. See, for example, Roth, Bennett, and Woods 1987.

45. Weick 1979. Similar ideas are increasingly current in the human-factors, man-machine interface literature.

46. James Reason refers to the embedding of representational errors as a form of "resident pathogen" (Reason 1989).

47. For example, nuclear power plant operators must balance their interest in taking a plant down when they suspect something is not quite right against the cost to the utility of the loss of power.

48. Cf. Rasmussen 1989; Reason 1989; Perrow 1986; Weick 1988. One corollary of the system-embedded–error hypothesis is that this class of accident is a growing fraction of all accidents. Since the public and many administrators and regulators (not to mention engineers) are demonstrably unwilling to accept the relative growth of this subset, such accidents are more often than not assigned to the category of "human error," even though, in reality, they should be assigned to a new class of systemic technical failures over which real operators, in real situations, have only the illusion that they are in control of events.

49. Rochlin, La Porte, and Roberts 1987.

50. Note that at times the *Vincennes* itself may serve as anti-air warfare coordinator for a larger group of ships, and therefore is at least as fully equipped (and probably with more modern equipment) as a carrier.

51. Roberts and Rousseau 1989. See also Rochlin, La Porte, and Roberts 1987.

52. Rochlin 1989.

53. For a comparable case study with similar conclusions, see Lanir 1989.

54. Weick (forthcoming). For a discussion in the context of the U.S. Navy and the increasing complexity of its equipment, see Johns 1988.

55. Blair 1985 is a superb analysis of the status and quality of U.S. nuclear weapons command and control, discussing both positive and "negative" control modes.

REFERENCES

Bellin, David, and Chapman, Gary. 1987. *Computers in Battle: Will They Work?* New York: Harcourt Brace Jovanovich.

Blair, Bruce G. 1985. *Strategic Command and Control.* Washington, D.C.: Brookings.

Byron, Capt. John L. 1987. "The Surface Navy is Not Ready." *Proceedings of the Naval Institute* 113/12/1018 (December), pp. 34–40.

Canan, James W. 1988. "Sorting Out the AirLand Partnership." *Air Force* (April), pp. 50–59.

Deitchman, Seymour J. 1983. *Military Power and the Advance of Technology: General Purpose Military Forces for the 1980s and Beyond.* Boulder: Westview Press.

Din, Allan M., ed. 1987. *Arms and Artificial Intelligence: Weapons and Arms Control Applications of Advanced Computing.* New York: Oxford University Press.

Gibson, James William. 1988. *The Perfect War.* New York: Vintage.

Johns, Eric. 1988. "Perfect is the Enemy of Good Enough." *Proceedings of the Naval Institute* (October), pp. 37–48.

Keegan, John. 1987. *The Mask of Command.* New York: Viking.

Landau, Martin, and Stout, Russel, Jr. "To Manage is Not to Control: Or the Folly of Type II Errors." *Public Administration Review* 39 (March/April 1979): 148–56.

Lanir, Zvi. 1989. "The Reasonable Choice of Disaster—The Shooting Down of the Libyan Airliner on 21 February 1973." *Journal of Strategic Studies* 12, 4 (December): 479–93.

Lanir, Zvi; Fischoff, Baruch; and Johnson, Stephen. 1988. "Military Risk-Taking: C3I and the Cognitive Function of Boldness in War." *Journal of Strategic Studies* 11, 1 (March): 96–114.

La Porte, Todd R. 1982. "On the Design and Management of Nearly Error-Free Organizational Control Systems." In *The Accident at Three Mile Island: The Human Dimensions.* Boulder: Westview Press.

March, J. G., and Weissinger-Baylon, R., eds. 1986. *The Control Revolution.* Cambridge, Mass.: Harvard University Press.

O'Connell, Robert. 1989. *Of Arms and Men.* New York: Oxford.

Perrow, Charles. 1984. *Normal Accidents: Living with High Risk Technologies.* New York: Basic Books.

————. 1986. *Complex Organizations: A Critical Essay.* 3d ed. New York: Random House.

Rasmussen, Jens. 1986. *Information Processing and Human-Machine Interaction: An Approach to Cognitive Engineering.* Amsterdam: North Holland.

————. 1989. "Human Error and the Problem of Causality in Analysis of Accidents." Invited paper for the Royal Society meeting on Human Factors in High Risk Situations, London, 28–29 June.

Rasmussen, J., and Batstone, R., eds. Forthcoming. *Safety Control and Risk Management.* Proceedings of a conference in Washington, D.C., October 1988. New York: Wiley.

Reason, James. 1989. *Human Error: Causes and Consequences.* New York: Cambridge University Press.

Roberts, Karlene H. Forthcoming. "Bishop Rock Dead Ahead: The Grounding of the USS

Enterprise." *Proceedings of the Naval Institute.*

Roberts, Karlene H., and Rousseau, Denise. 1989. "Research In Nearly Failure-Free, High-Reliability Systems: 'Having the Bubble.'" *IEEE Transactions* 36: 132–39.

Rochlin, Gene I. 1989. "Informal Organizational Networking as a Crisis-Avoidance Strategy: U.S. Naval Flight Operations as a Case Study." *Industrial Crisis Quarterly* 3, 2: 159–76.

Rochlin, Gene I.; La Porte, Todd R.; and Roberts, Karlene H. 1987. "The Self-Designing High-Reliability Organization: Aircraft Carrier Flight Operations at Sea." *Naval War College Review* (Autumn), pp. 76–90.

Roth, E. M.; Bennett, K. B.; and Woods, D. D. 1987. "Human Interaction with an 'Intelligent' Machine." *International Journal of Man-Machine Studies* 27: 479–525.

Sarkesian, Sam C., ed. 1980. *Combat Effectiveness: Cohesion, Stress, and the Volunteer Military.* Sage Research Progress Series on War, Revolution, and Peacekeeping 9. Beverly Hills: Sage.

Squires, Sally. 1988. "The 'Glass Cockpit Syndrome.'" *Washington Post*, 11 October.

Thompson, James D. 1967. *Organizations in Action.* New York: McGraw-Hill.

U.S. Department of Defense. 1988. *Report of the Formal Investigation into the Circumstances Surrounding the Downing of Iran Air Flight 655 on 3 July 1988.* Washington, D.C.: DOD. 28 July.

U.S. General Accounting Office. 1988. *Weapons Testing: Quality of DOD Operational Testing and Reporting.* Washington, D.C.: USGAO.

U.S. House of Representatives Committee on Armed Services. 1987. "Report on the Staff Investigation into the Iraqi Attack on the USS *Stark*." 14 June.

van Creveld, Martin. 1985. *Command in War.* Cambridge, Mass.: Harvard University Press.

————. 1989. *Technology and War.* New York: The Free Press.

Vlahos, Michael. 1988. "The *Stark* Report." *Proceedings of the Naval Institute* (May), pp. 64–67.

Weick, Karl E. 1976. "Educational Organizations: Loosely Coupled Systems." *Administrative Science Quarterly* 21 (March): 1–19.

————. 1979. *The Social Psychology of Organizing.* 2d ed. Reading, Mass.: Addison-Wesley.

————. 1988. "Technology as Equivoque: Sense-Making in New Technologies." Paper prepared for a conference on Technology, Carnegie-Mellon University, 28–30 August 1988.

————. Forthcoming. "Mental Models of High-Reliability Systems." *Industrial Crisis Quarterly*..

Wilson, George C. 1986. *SuperCarrier.* New York: MacMillan.

————. 1989. "The Risks of Shooting First." *Washington Post*, 28 January.

THE AVIATION SAFETY COMMISSION: A TASK FORCE APPROACH TO THE EVALUATION OF LARGE TECHNICAL SYSTEMS

Clinton V. Oster, Jr.
School of Public and Environmental Affairs
Indiana University

Introduction

In 1986, Congress created an independent presidential commission to study aviation safety and make recommendations to Congress and the president by 18 April 1988. The Aviation Safety Commission was to study a large technical system in a politically charged environment.

An Overview of Aviation Safety

A safe commercial airline flight requires a set of complex systems to work together. The aircraft must be designed for a wide variety of flight and environmental conditions, must be built according to the design specifications, and must be maintained to continue the designed level of performance. Flight and ground crews must be trained to operate each specific type of aircraft in a variety of weather conditions and to respond quickly and properly to equipment failures and other unexpected events. The air traffic control system, in coordination with the flight crew, must prevent aircraft from colliding with one another and with other objects while the aircraft are on the ground and in the air. Each of these systems must be able to maintain flight safety in the face of both mechanical malfunction and human error.

Aircraft design, of course, involves the airframe and a whole series of systems including propulsion, flight control, communications, navigation, cabin environmental control, and steering and braking for ground movement. Typically these systems are designed and manufactured by many different companies, each with its own practices and incentives, yet all must operate smoothly together.

Aircraft designers must make tradeoffs among the conflicting demands of producing an aircraft that is easy and inexpensive to construct, inspect, maintain, and operate. To be a commercial success, the aircraft must also perform a wide variety of missions in terms of passenger loads, flight lengths, and operating environments with a high degree of schedule reliability. Both unnecessary cost resulting from overdesign and inadequate reliability from underdesign can lead to commercial disaster.

Before an aircraft can be put in service, it must be certified by the Federal Aviation Administration (FAA) as safely designed. If the aircraft is to be operated abroad, it must be certified by the FAA's counterparts in those countries. While there is considerable

T. R. La Porte (ed.), Social Responses to Large Technical Systems, 127–141.

international cooperation, aircraft certified in one country may not be accepted in another, particularly if the aircraft has unusual design features.

The certification process adds to the cost of bringing a new technology to market. Not only must the manufacturer convince the customers that the new technology is safe, reliable, and cost-effective, government agencies throughout the world must also be convinced of the technology's safety over the life of the aircraft. The government's role thus alters the tradeoff between standardization and technical progress and the question of when a new technology is worth introducing.

Once a design is approved, assuring manufacturing quality control is complicated by many different component manufacturers spread across diverse geographic locations. An aircraft "built" by Boeing or McDonnell Douglas is, in fact, only assembled by that company from components manufactured throughout the world. Shortcomings in design or manufacturing may not become evident until after the aircraft has been in service for years. Extraordinarily detailed records must be kept during each aircraft's manufacture and life to allow lessons learned from problems with one aircraft to be applied quickly to others.

For example, in the first few days following the 1989 crash of a United Airline's DC-10 in Sioux City, Iowa, the failure of a particular engine component was suspected of initiating the sequence of events that ultimately led to the accident. One concern was that the batch of metal alloy from which this component was made might have been defective and that other components made from the same batch might also be subject to failure. The record-keeping system is such that within a few days, all the other components made from that particular batch had been located and inspected and, as it turned out in this case, found to be free of defects.

An aircraft designed and built to be safe can quickly become unsafe without proper maintenance. Maintenance needs vary by component. Some need to be serviced or replaced based on the amount of time since they were installed; others based on the number of hours in flight; others based on the number of takeoffs and landings; and still others are left in service until an inspection reveals the need for replacement. Frequently, components can fail before they are scheduled for service or replacement. In anticipation of such failures, aircraft contain backup systems for virtually all components necessary for safe flight. Because such redundancy allows an aircraft to be flown with some inoperative components, redundancy also allows the possibility of some discretion in maintenance. How much redundancy is needed for a particular flight? Should an aircraft with an inoperative component for which there is a backup be grounded immediately until a repair is made or can it be flown until the end of the day or until it reaches a maintenance facility? The government has taken on an oversight role and limited the range of discretion a carrier has in making these decisions.

This oversight role is complicated by markedly different approaches that are equally safe and effective. For example, one carrier may choose to replace components on a regular schedule regardless of the condition of the component. Another carrier may choose to leave a component in service longer and ensure its performance by frequent inspection and testing. Regardless of approach, detailed records must be kept of maintenance practices to help assess the effectiveness of particular maintenance programs.

An aircraft that is properly designed, built, and maintained, must still be operated by skilled pilots working together as a team. The flight crew must have been trained to fly the aircraft safely under a wide variety of circumstances and their flying skills must be kept sharp by recurrent training. The advent and increasing sophistication of flight simulators have improved flying skills training dramatically in recent years. Questions arise, of course, about how much initial training is needed and how often and how extensive recurrent training should be. Maintaining the proficiency of individual pilots is not enough to ensure safe flight. Smooth coordination among crew members is crucial, particularly in emergencies. Allowing flight crews to stay together as units for a period of time might well aid crew coordination. However, it would make efficient crew scheduling more difficult and it would hamper an airline's ability to give pilots increased responsibilities with greater experience. Similarly, having a pilot specialize in only one type of aircraft might enhance safety, but again at the expense of schedule flexibility and skill progression.

The final major link in a safe flight is the air traffic control system. The design, construction, maintenance, and operation of the air traffic control system has many of the same elements as the design, construction, maintenance, and operation of aircraft. Designers of air traffic control equipment, be it radar, computers, or communications equipment, face the same pressures for standardization in the face of technological progress. Design is often constrained by the need for equipment at one site to be compatible both with equipment elsewhere in the air traffic control system, in the United States and abroad, and with equipment in the large and varied fleet of aircraft to be controlled. As with an aircraft, the air traffic control system must incorporate redundancy to guard against both component failure and human error. A well-designed, constructed, and maintained system must still be operated skillfully. As with aircraft crew, air traffic control operators, be they controllers or electronics maintenance technicians, must both be individually proficient and must work together smoothly as a team. Both initial and recurrent training are critical and personnel must be given the opportunity to move to increasingly demanding responsibilities as skill levels and experience increase.

The Aviation Safety Commission

In the mid-1980s, a series of well-publicized aviation accidents heightened concern by the public, the media, and the government that there might be some slippage in the excellent safety record the U.S. airline industry had amassed over its sixty-year history. The airlines' economic environment had changed dramatically in the late 1970s with the advent of deregulation. The subsequent growth in both the number of airlines and in airline travel had placed increased pressure on the airlines, the air traffic control system, and the FAA's inspection and certification programs. More recently, airline mergers placed pressure on the FAA to ensure that merging carriers combined often diverse approaches to training and maintenance without degradation of safety.

Against this backdrop, Congress, at the urging of Senator Byrd of West Virginia, passed the Aviation Safety Commission Act of 1986 (P.L. 99-591) creating an independent presidential commission to examine aviation safety. The Aviation Safety Commission was given eighteen months following the passage of the Act to study aviation safety and report back to Congress and to the administration. The commission's charter, shown in Exhibit 1, provided the charge to the commission and the questions to be addressed. These questions address two fundamental issues: (1) Has airline deregulation harmed airline safety?; and (2) Is the safety role of the federal government organized properly?

The Aviation Safety Commission was to be composed of seven commissioners, all appointed by the president without congressional confirmation. Four of the seven commissioners, including the chairman, were required to have no current or past ties to the aviation industry. The commissioners were to have their expenses covered but were not to receive compensation for their service. The executive director of the commission was also to have no current or past ties to the aviation industry. In essence, the commission was structured from the start to have part-time volunteer commissioners, a majority of whom were to have no background or experience in aviation. The head of the staff, although full-time, was similarly to have no background in aviation. The commission was created by a Congress dominated by democrats, but the commissioners were all to be appointed by a Republican president.

This arrangement placed President Reagan in a difficult situation. A centerpiece of the commission's charge was to assess the impact of airline deregulation on safety. Airline deregulation was passed by Congress in 1978 with bipartisan support and was signed into law by a Democratic president. While President Reagan continued to support deregulation strongly, Senator Byrd had come to oppose it and was an advocate of re-regulating the industry. A finding by a Republican-appointed commission that deregulation had harmed safety would be a strong asset in Senator Byrd's re-regulation efforts. A commission finding that deregulation had not harmed safety could easily be dismissed as a partisan whitewash by a group with little expertise. In retrospect, it is not surprising that the White House did not appear eager to appoint this commission. Indeed, the last of the commissioners was not appointed until June 1987, no staff were added until July, and the enabling legislation required that the commission make its report on 18 April 1988. No extensions to this deadline could be made without congressional action and Senator Byrd's staff made it clear that an extension was unlikely.

Assembling the Aviation Safety Commission

Four of the commissioners clearly fulfilled the legislative requirement of having no ties to the aviation industry. The chairman, Dr. John Albertine, was trained as an economist, had some experience as a Washington lobbyist, and was then vice-chairman of Farley Industries, a privately held company based in Chicago whose primary holding was Fruit of the Loom, a garment-making company. A second commissioner was John Byington, a Washington-based attorney who had previously headed the Consumer Products Safety

Commission. A third commissioner was Professor Joseph Kalt, an economics professor at Harvard's Kennedy School of Government whose specialty was energy and natural resource economics and regulation. The fourth commissioner was Thomas Wathen, owner of California Plant Protection (now CPP-Pinkerton), a private security firm. While the first three had absolutely no experience in aviation, Mr. Wathen was a private pilot whose hobby was antique airplanes.

Three of the commissioners were permitted to have aviation ties, but none of those appointed in this category had engineering or technical expertise. Michael Levine was a lawyer by training and a professor at the Yale School of Management. He had previously been a staff member at the Civil Aeronautics Board and later president of New York Air. He remained a consultant to Texas Air Corporation. Russell Meyer was also a lawyer by training and was chairman of Cessna Corporation and an active pilot. John Robson was also a lawyer by training and was dean of the business school at Emory University. He had previously been chairman of the Civil Aeronautics Board prior to deregulation and was a member of the board of directors of Continental Airlines.

The seven commissioners were mostly Republican, mostly conservative, and mostly supporters of the free market. To the extent that they had a prior opinion on airline deregulation, they were in favor of it. To the extent that they had opinions on aviation safety, they did not believe there was a crisis at hand, although some believed safety could easily be improved.

Upon appointment of John Albertine as chairman, the commission was complete and Senator Byrd sent a letter to the chairman clarifying that which he intended the commission to examine. The charge to the commission, as stated in Senator Byrd's letter is summarized in Exhibit 2. Senator Byrd's letter raised questions ranging from those as broad as in the legislation to several very specific issues.

Four issues were raised in the senator's letter that had not been mentioned in the legislation. The first was the safety of commuter carriers in providing service to small communities. This issue was of particular concern to the senator because at that time West Virginia received no scheduled jet service and was served only by commuter airlines. The second was the impact of President Reagan's dismissal of 11,500 striking air traffic controllers in August 1981. Several of the senator's colleagues in Congress strongly supported rehiring these controllers. The president, not surprisingly, opposed rehiring the controllers. The third new issue was the risk general aviation aircraft might pose to the safety of commercial airline service and what steps might appropriately be taken to reduce that risk. The fourth issue was whether money in the Airport and Airways Trust Fund was being spent wisely and quickly enough to modernize the air traffic control system. These latter three issues were ones on which many people in Washington and in the aviation community had already formed hardened battle lines.

With a part-time commission, the full-time staff takes on most of the work that forms the basis and support for findings and recommendations. With the majority of commissioners with no experience in the area, the staff may also help shape the direction of the efforts. Normally, the staff is assembled and managed by an executive director. For the Aviation Safety Commission, however, the executive director was required to have no current or past ties to the aviation industry. The commission appointed an executive director with no aviation experience, but with considerable experience in the ways of Washington. A research director was also appointed with experience in aviation safety research and research management. The research director's experience, however, was largely confined to the academic community and he had little experience with the aviation industry. The enabling legislation made no mention of a research director, so no restrictions were placed on that appointment. The Aviation Safety Commission's unusual appointment of both an executive director and a research director proved confusing to many in Washington, but did not pose problems for the commission's internal workings.

Assembling a staff for a commission such as this was difficult. By the time the commission was appointed, the time remaining for the commission's work was less than a year and there was no possibility of continued employment with the commission after completion of the final report. This short duration also meant that it was unlikely that anyone could be convinced to relocate to the Washington area just to participate in the project. While a great many people applied for staff positions, very few had either valuable experience or useful skills to offer. Many of the applicants appeared to see the commission as a means of turning around a faltering career. The available pool of talent was quite limited. The staff selected for the commission were generally young and inexperienced, but talented and eager to work hard.

Many consultants also offered their services to the commission. Here the principal problem was one of time. With less than a year until the report was due and with part of the time needed to define which of the many possible issues the commissioners wished to pursue, there simply was not sufficient time to define clearly identifiable projects and have them completed by specialized consulting firms. The firms and consultants with proven track records also had limited ability to respond to the commission's needs on such short notice because of their existing commitments. The consultants who did most of the work for the commission were those with both broadly based skills and experience and with prior experience working with either the research director or one of the commissioners. One notable exception to this pattern was a recently retired FAA employee whose work proved particularly helpful on one topic.

The Commission's Work Plan

The commission began its work in July with two days of hearings in Washington. The hearings provided few surprises and were similar to previous congressional hearings; most of people who testified had testified in the past on the same topics saying much the same thing to one or more congressional committees. As is common with such hearings, few of the witnesses lived more than thirty miles from Washington. The hearings were valuable in introducing the commissioners and staff to some of the issues and some of the positions various groups held on these issues. They also provided an opportunity for the commissioners and staff to begin to develop working relationships.

The hearings also announced to the aviation community that the Aviation Safety Commission had finally begun its work. Following the hearings, a flood of information and offers to help the commission with its work began to descend on the commission's headquarters. Some information came from those who testified at the hearings and provided background and amplification of their testimony. Many in the aviation community were concerned that the commission's lack of experience and expertise in aviation might lead to recommendations based on misconceptions of the problems. The sources of information and offers can be grouped as follows:

Trade Associations. The most aggressive groups were the Washington-based trade associations representing various segments of the aviation industry. With varying degrees of subtlety, persistence, and focus, each wanted to bring a perceived problem and a proposed solution to the commission's attention. Some were helpful in providing useful information. Others seemed bent on insulting the intelligence of even the most experienced staff members.

Individual Airlines. In addition to the offers of help by the airlines' trade association, individual airlines offered help as well. The offers were primarily invitations to visit maintenance and training facilities to gain a better understanding of the regulatory process from the airline's perspective. None of the airlines refused requests to visit maintenance and training facilities. These visits proved helpful in gaining a broader perspective on the regulatory process. Similarly, airlines allowed commissioners and senior staff to ride in the cockpits during airline flights to observe procedures and talk with crew. These rides also helped the commission gain a broader perspective on the FAA inspections and air traffic control procedures.

Airframe Manufacturers. As with the individual airlines, offers from manufacturers were primarily invitations to visit facilities to gain a better understanding of the manufacturer's perspective on the regulatory process. As with airlines, all requests for visits were honored.

Equipment Vendors. In several cases, vendors offered the commission their insights into specific aviation safety hazards. These insights invariably included solutions involving their products.

Labor Unions. Unions representing the pilots, machinists, air traffic controllers, and airways facilities technicians were all interested in providing information to the commission.

In addition to those who came to the commission with information, others offered to provide whatever information or assistance the commission requested. The Federal Aviation Administration, in particular, offered full cooperation. The new FAA Administrator, Alan McArter, had made his first public appearance following his confirmation at the commission's hearing. In his testimony, he pledged the full cooperation of his agency and, during the commission's study, he made good on that pledge. Not only were FAA personnel in the Washington headquarters cooperative, but the cooperation extended to the personnel in the field, where the views were often quite different from those in Washington.

The basic approach taken by the commission in its work was straightforward. Following the July hearings, the staff spent two months developing a briefing book built around a list of questions the commission was charged to address, for the commissioners in preparation for an October meeting. The principal sources of these questions were the enabling legislation and the letter from Senator Byrd. Additional questions also emerged from the July hearings and subsequent staff interviews with aviation safety experts. Similar questions addressing the same issue were grouped together and related issues were grouped into major topic areas. Exhibit 3 presents the structure of the briefing book outlining the four major areas and the twenty-nine separate briefing papers.

For each of these briefing papers, background material was assembled and summarized to explain the significance of the issue, a research plan for exploring and developing recommendations was presented, and a staff member and one or more consultants were identified to conduct the research. Each research plan outlined the necessary activities for the staff and consultants and recommended activities whereby the commissioners could gain first-hand knowledge of the issue. The briefing papers were based largely on published material, the commission hearings, and staff interviews conducted within the Washington area.

Several things became apparent at this point. First, several of the issues placed before the commission were easily resolved. For example, the commission charter raised the issue of whether the National Transportation Safety Board's (NTSB) recommendations should be made mandatory (Exhibit 1). Not surprisingly, the Federal Aviation Administration was opposed to this idea. Perhaps more surprisingly, the NTSB was also opposed to the idea. Indeed, no one could be found who advocated this position. Similarly, Senator Byrd's letter raised the question of whether the FAA should return to the air traffic control practice of airborne holding in response to delays instead of the current practice of holding aircraft on the ground (Exhibit 2). Again, opposition to airborne holding was strong and no one could be found who advocated it.

Second, the major issue before the commission—whether deregulation had resulted in a degradation in safety—appeared headed quickly toward a resolution that the commissioners had favored from the beginning. The commissioners had been appointed by staunch advocates of airline deregulation. The empirical evidence was emerging to demonstrate that the U.S. airlines had amassed a better safety record in the years following deregulation than during any period prior to deregulation. Thus, the commissioners were not likely to be drawn into a partisan defense of economic deregulation, but could instead focus on developing recommendations to improve an already safe system.

Third, even with some of these issues resolved, there simply was not enough time to explore all of the remaining issues placed before the commission in sufficient depth to make sensible recommendations.

Fourth, and perhaps most important, for some of the issues there was no amount of research by the staff that would make the commissioners themselves comfortable in making recommendations. Because of their backgrounds, the commissioners were far more eager to undertake some issues than others. Even with technical expertise at their disposal, the commissioners were reluctant to delve into policy disagreements involving technical issues.

At the October meeting, the research areas to be given emphasis were selected and the research questions refined. The staff was directed to develop "Issues and Options" papers for each of the four areas. Each of these papers was to provide analyses of the issues in the area and was to identify potential recommendations for the commission to make. A meeting to discuss the completed Issues and Options papers and decide on preliminary recommendations was scheduled for mid-January. Exhibit 4 outlines the structure of the Issues and Options Briefing Book.

At this point, staff work began to expand outside the Washington area as a series of trips and field interviews began. Reactions to these trips was highly favorable in the field, but mixed in Washington. Some in Washington applauded the approach, others felt it was unnecessary, while still others seemed troubled by it. Several of the eventual recommendations were strongly influenced by information gained on these trips. Many of the issues examined had taken on a life in Washington that differed markedly from the perspective in the field.

The general thrust of the commission's recommendations was decided at the January meetings. The intent was that the next two months would be spent refining the details of the recommendations and writing the final report. The commissioners were strongly in agreement that they wished to develop unanimous recommendations. They felt that a minority report and recommendations would weaken the impact of the entire effort.

Up until this point, the commission's work had not progressed markedly differently than the early and middle stages of other multidisciplinary applied research projects. The lack of technical expertise and experience in the topic areas on the part of some of the commissioners had mixed effects. Initially, inexperience was a drawback because of the time and effort needed to bring some of the commissioners up to speed. However, it's easy to make too much of this. Much of this background work would have to have been done eventually anyway. If the commission's work was to have any impact, it had to be both easily understood and convincing to a non-expert audience—Congress and the Executive Branch. The time spent educating the commissioners also developed the necessary background material to write a report that could be understood by its intended audience.

The lack of expertise also turned out to have some benefits. Because a majority of the commissioners were new to the topic, they were not well versed in the folklore about what the important problems, and their underlying causes, were. As a result, the commissioners tended to ask "Why is this important?" and "How big a problem is this really?" about almost everything. The result was that some of the issues that had taken on the aura of a holy war in the aviation community were found by the commission to be entertaining, but not terribly important. Rather than focus on each problem separately, the commission tried to look for the underlying causes of groups of problems.

Developing Recommendations in a Political Environment

The Aviation Safety Commission was a creature of Washington and had to develop its recommendations in a political environment. The commission's recommendations did not become public until the release of the final report on 18 April 1988. These recommendations were outlined in the executive summary of the final report, which is contained in Exhibit 5.

From the beginning of the commission's life, there were widely divergent expectations about what the commission would recommend. Some observers expected a benign endorsement of the aviation system as safe and needing at most a little tinkering around the edges. Others expected the commission to produce a resounding call for the privatization of at least the air traffic control function and perhaps other of the FAA's functions as well. Still others expected that the commission would find itself confronted with the inescapable conclusion that the margin of safety had degraded in the wake of deregulation and that some backing away from economic deregulation was the only logical step. The problem was that these differing expectations were held by members of Congress, congressional staff members, and high level executive branch officials.

As the commission's recommendations began to take shape following the January meeting, the thrust of these recommendations was discussed privately with congressional staff and officials in the Department of Transportation (DOT), the Office of Management and Budget (OMB), and the White House. One way to characterize the commission's view was that while safety had not degraded in the wake of deregulation, there were many ways in which it could be improved that had not been implemented. Furthermore, there was enough blame for this lack of implementation to be shared by FAA, DOT, OMB, and Congress. As a result, the commission's recommendations had something to make almost everybody mad.

From early February until the commission's report went to press in early April, attempts were made by some of the interested parties to convince the commission to change some of its recommendations while retaining others. Of course, which recommendations were to be changed and which were to be retained varied with the party. Similarly, the techniques used to attempt to convince the commission varied with the party and with the proximity of the date of the commission's report.

Aviation safety recommendations are the sort of thing about which well-intentioned reasonable people can disagree. The commission's recommendations represented several carefully crafted compromises to accommodate differing views among the commissioners. The pressure from interested parties to change recommendations placed considerable strain on these compromises. Moreover, several of the commissioners may have felt they had more to lose by angering various people than did others on the commission. In the end, despite all of the pressure, the commission's final recommendations did not differ in any substantial way from those agreed upon in January.

It is perhaps ironic that most of the commissioners' time in the meetings from January on and most of the pressure in various directions centered around the recommendations on removing the FAA from the DOT, making it user-funded, and removing the OMB from the regulatory oversight process. For the most part, the commission recognized in January that these recommendations had only the smallest chance of actual implementation. The problems inherent in the relationship among the FAA, the Office of the Secretary in the DOT, the OMB, and Congress were judged sufficiently important, however, to warrant a dramatic recommendation that would command attention. The commission's other recommendations were largely overlooked by the media and most have quietly been implemented by the FAA.

It is difficult to make a final judgment on the success of the Aviation Safety Commission in confronting a large technical system such as aviation safety because it is difficult to know what constitutes success for such a commission. The non-technical background of the commissioners prevented the commission from addressing the technical aspects of aviation safety or resolving technological arguments. But the commissioners also avoided becoming absorbed in narrowly focused technical issues and being drawn in by some of the widely held myths about what the important issues were. As a group, the commissioners went to their strengths and backgrounds in choosing issues with which they were comfortable making recommendations. Perhaps because of this, most of their recommendations have been implemented and all the issues they identified have been addressed in some manner. It remains to be seen, however, whether aviation safety will improve as a result.

EXHIBIT 1

The Charge as Stated in the Charter

Has airline deregulation had an adverse effect on the margin of aviation safety?

Is the practice of airline self-compliance with respect to aviation maintenance standards an outmoded approach in an environment designed to maximize cost-savings?

Has the FAA adequately used its resources to ensure aviation safety?

What measures should be taken to improve the enforcement of federal regulations relating to aviation safety?

Are the dual responsibilities of the FAA of promoting commercial aviation and ensuring aviation safety in conflict and does such conflict impede the effective maintenance and enhancement of aviation safety?

Should the FAA be reorganized as an independent federal agency with the promotion, maintenance, and enhancement of aviation safety as the sole responsibility of such agency?

Should the promotion of commercial aviation be assigned as a responsibility to another agency of the federal Government?

What is the most appropriate and effective organizational approach to ensuring aviation safety?

Is it feasible to make mandatory certain or all of the safety recommendations issued by the National Transportation Safety Board?

EXHIBIT 2

The Charge as Stated by Senator Byrd

What has been the impact of airline deregulation on the margin of safety of our air traffic system?

Given the substantial growth in the number of commuter airlines and the concomitant increase in commuter traffic, should the FAA require a more stringent minimum standard for commuter pilot training and for commuter airline maintenance practices?

Is the Federal Aviation Administration properly structured to provide effective management of the air traffic system?

Has the combination of steadily increasing traffic coupled with the firing of 11,500 controllers left the air traffic system overburdened and understaffed?

What are the minimum staffing standards necessary to ensure that the air traffic system is managed safely?

Should the flow of air traffic be restricted so that the volume of that traffic is consistent with the capability of the controller force at each Air Traffic Control Center?

Should the FAA be reducing aircraft separation standards, normalizing flow controls, and returning to the practice of airborne holding in order to minimize delays on the ground?

What can be done to accelerate the development of collision avoidance systems for large commercial airliners?

What can be done to accelerate the modernization of the facilities and equipment in our air traffic system?

Why has the implementation of the National Air Space (NAS) Plan fallen six to eight years behind schedule?

Do we need additional controls on general aviation aircraft?

How can we ensure that small general aviation aircraft will more safely coexist with large commercial airliners?

Should transponders be required on all general aviation aircraft?

Are the dual responsibilities of the FAA to both promote air commerce and to regulate it in conflict?

Should the FAA once again be made an independent agency?

EXHIBIT 3

October Briefing Papers

Area 1: Deregulation and Safety

The Margin of Safety
Safety Record and Measures of Safety
Impact of Deregulation and Commuter Carriers
Financial Impacts on Safety

Area 2: FAA Operation, Administration, and Internal Structure

FAA Internal Management Structure
Role of FAA Regional Offices
Maintenance Inspections
FAA Rulemaking and Standards
A Fixed Term for FAA Administrator
FAA Employee Relations
FAA Staffing Levels

Area 3: The Air Traffic Control System

Collision Avoidance Systems
National Airspace System Plan Hardware
Air Traffic Control Operations Research Solutions
Air Traffic Control Demand Management Solutions
Airport Expansion
Airport Operations Research Solutions
Airport Demand Management
Organizational Status of Air Traffic Control
General Aviation in Congested Airspace
Altitude Encoding Transponders

Area 4: External Placement of the FAA

FAA's Dual Mandate
The Airport and Airways Trust Fund
FAA/DOT Relationship
External Placement of the FAA
NTSB Recommendations

Additional Briefing Papers

National Airspace System Plan
Commercial Airline Margin of Safety
FAA Near–Midair Collision Reports

EXHIBIT 4

January Issues and Options Papers

Area	*Title*
1	Deregulation and Safety
2	FAA Safety Inspection Programs
3	General Aviation in the Air Traffic Control System
4	Organizational Approach to Ensuring Aviation Safety

Additional Supporting Papers

5	*The Government Corporation*
6	*Organizational Status of the Air Traffic Control System*
7	*List of People Interviewed*

EXHIBIT 5

Executive Summary of the Final Report

After months of study, hearings, meetings around the country, and staff reports, the Aviation Safety Commission unanimously concludes that the nation's air transportation system is safe. However, safety is being maintained to an increasing extent through delays and other inconveniences.

Air transportation has changed during the past decade. Economic regulations that had shaped the industry since the 1920s were replaced by airline deregulation. The resulting increased competition has lowered fares, expanded service, and brought air travel to millions who had not previously been able to afford to fly. It has also made the FAA's job much more difficult.

The Aviation Safety Commission concludes that the present safety regulatory structure designed to ensure aviation safety is inadequate to deal with future growth and technological change. Now is the time to equip the regulatory system to accommodate changes in the numbers and kinds of aircraft, to take advantage of new technology in aircraft design and manufacture, to respond to heightened sensitivity on the part of the public to aviation safety, and to act on the backlog of potentially worthwhile safety improvements that have been languishing because of diffused authority and accountability. In short, now is the time for decisive action by Congress and the executive branch.

The Aviation Safety Commission believes that the federal government must continue to play the central role in ensuring safe operation of the U.S. aviation system. We also share the common perception that, while the system is safe for now, the present governmental structure is not working effectively enough to ensure its safety in the future. Therefore, we agree unanimously that a major structural overhaul is essential. We believe that the regulatory process must remain governmental in character and should not be taken out of the federal government or removed from public accountability.

The commission's recommendations address in a constructive way all of the issues which have been raised in the current debate and reflect the input we have received from the Secretary of Transportation, the FAA, former FAA Administrators, members of Congress, the NTSB, consumers, and industry experts.

Specifically, the Aviation Safety Commission recommends that FAA be transferred from the Department of Transportation and be established as a user-funded authority which is:

- overseen by a nine-member Board of Governors appointed by the President and confirmed by the Senate;
- managed by an Administrator who is appointed and confirmed for a term of seven years;
- subject to agency-wide oversight by a Director of Aviation Safety who is appointed and confirmed for a term of seven years;
- freed from the constraints of the federal civil service and procurement systems.

The Director of Aviation Safety has the authority to initiate rulemaking as well as disapprove regulations promulgated by the Administrator, and also has the authority to enforce compliance by the Administrator of existing rules and regulations. Decisions by the Administrator and the Director of Aviation Safety are appealable to a Safety Committee of the Board of Governors composed of the Administrator, the Director of Aviation Safety, the Secretary of Transportation, the Secretary of Defense, and a public member, and hence are not subjected to OMB review.

The Aviation Safety Commission also recommends the following agenda for improving aviation safety:

Safety Inspection Programs

- national rather than regional certification programs for major and national jet carriers;
- establishment of a nationwide inspection program for all size carriers with a combination of regular, in-depth, and surprise inspections;
- separation of certification and surveillance functions in the new Authority;
- priority inspections for carriers undergoing major change;
- increasing the inspector workforce to accommodate these changes.

Regional Airline Safety

- reducing differences in equipment standards between regional and national carriers, with all aircraft providing scheduled service being required eventually to meet Air Transport Category Aircraft (Part 25) standards;
- reducing differences in operating practices between regional and national carriers, with all carriers eventually being required to meet Part 121 operations requirements.

General Aviation in the Air Traffic Control System

- requiring all aircraft to be equipped with a Mode C transponder in buffer zones around all large, medium, and small hubs;
- stronger enforcement against buffer zone violators with a separate radar position dedicated to tracking and notifying violators in each buffer zone.

FAA Rulemaking

- process must be streamlined and restructured to include clear and unambiguous responsibility and accountability.

Airport Safety and Capacity

- base airport certification on passenger volume rather than type of equipment;
- review of existing policies and requirements with particular emphasis on signage, directional indicators, and taxiway and intersection markings.

Use of Operations Research

- need to enhance operations research capabilities for better utilization in problem solving.

HOW DO WE TREAT TECHNICAL UNCERTAINTY IN SYSTEMS FAILURE? THE CASE OF THE SPACE SHUTTLE *CHALLENGER* [1]

Trevor J. Pinch
Department of Sociology, University of York

Introduction

How certain are we that we understand the working of a technology? In this paper I attempt to raise some of the issues posed by technical uncertainty for social scientists in their treatment of large technical systems. Such issues are always present, but they become most salient when systems undergo failure. As a way into the discussion, I will focus initially upon Charles Perrow's treatment of the topic in his book *Normal Accidents: Living with High Risk Technologies*.. I view Perrow's work through the perspective of sociology of science—a discipline which is familiar with treating technical uncertainty in the shape of disputed knowledge claims about the natural world. It is argued that there are close parallels between uncertainty as encountered in the context of technology and as found in the natural sciences. The issue of technical uncertainty is further examined through the case of the space shuttle *Challenger* accident.

Normal Accidents

Perrow, like a number of social scientists recently interested in failures in technological systems (e.g., Turner 1978), does not restrict himself to analyzing the *impact* of system failure on society, but is concerned with how failure arises in the first place. In his marvelously prescient book, he distinguishes between different types of technological systems and the differing types of accidents which they can undergo. Perrow is particularly concerned with tightly coupled, complexly interactive systems. Tightly coupled systems are ones where there is very little slack between the components of the system. What happens in one part directly affects what happens in another—nuclear power being a good example. A complex system has a variety of branching paths and feedback loops. Again nuclear power is the most straightforward example he discusses.

Perrow's book is prescient because he draws particular attention to the risks posed by tightly coupled complexly interactive systems. Such systems are the breeding grounds for the classic system accident or "normal accident." This is an accident which is particularly hard to foresee in detail because it involves failures in several parts of the system which are often linked in complex and subtle ways. Three Mile Island was such a normal accident where a combination of seemingly unrelated events (such as faulty valves and indicators) conspired together to produce a major system accident where the core of the reactor became exposed and a hydrogen bubble exploded. Perrow forewarns

T. R. La Porte (ed.), Social Responses to Large Technical Systems, 143–158.

us of the dangers of such accidents recurring in nuclear power stations (his book was written prior to the Chernobyl incident and points specifically to the risk posed by Soviet reactors which lack containment buildings) and he emphasizes the problems posed by other complex, tightly coupled systems, such as space missions, where the role of the astronauts as "system hackers" is all but bypassed.

Challenger—A Totally Unexpected Accident

One category of accident Perrow distinguishes is that of the catastrophic accident. This is an accident which threatens the survival of a potentially large number of human beings—generally more than those involved in the immediate operation of the technological system where the accident takes place (first-party victims) or users of the system such as passengers (second-party victims). Chernobyl is an obvious example of such a catastrophic accident where both outsiders' (third-party victims) and future generations' (fourth-party victims) lives were affected. The *Challenger* accident was not, strictly speaking, a catastrophic accident because, tragic as that accident was, the only loss of life was that of the seven astronauts on board.[2]

Although the *Challenger* accident was not a catastrophic accident in Perrow's sense, it was a peculiarly pernicious accident in the widespread effects it produced.[3] The impact of the accident was all the greater because it was viewed by so many people, either directly at the Cape, on live TV, or later that day in news bulletins and in newspaper photographs. The normality of this supposedly routine flight was emphasized by the ethnic and gender balance of the crew, one of whom was a humble schoolteacher chosen from thousands of volunteers. Damaging as the accident was to the psyches of ordinary Americans who witnessed the tragedy (the impact has been compared with that of the assassination of John F. Kennedy), much greater was the long-term damage done to the credibility of the American space effort. NASA's "Can Do" attitude—the legacy of the Apollo program; its cultivation of a sycophantic press; and its seal of presidential approval all meant that the shuttle had become the symbol of American pride, power, and prowess in space.

This type of accident, because of the symbolic potency of the technology and the widespread belief that it is reliable, is probably unique. Of course, to a degree all accidents are unexpected or untoward events, but there is a difference between a technology which is perceived to be accident-prone and one which is taken to be fundamentally safe because of all the precautions taken to ensure safety. Although *Challenger* was a "normal accident," it was, nevertheless for most people, a totally unexpected accident.

Having briefly summarized some of the main themes of Perrow's book and introduced the *Challenger* accident, I will now delve in a little more depth into the notion of a "normal accident" and the type of problems it raises when viewed from a sociology of science/technology perspective.

Accident Scenarios

In order to introduce us to the notion of a "normal accident," Perrow begins his book by constructing an everyday scenario in which a person is preparing to leave home for an important job interview in a nearby town. A series of unforeseen events (such as the car keys getting locked in the apartment, a fault in a neighbor's car, a bus strike over a safety issue and the subsequent extra demand for cabs) ensue which prevents the candidate going to the interview. This then is a typical normal accident with a number of complex linked events, some being strongly coupled to each other, others being weakly coupled, and with back-up systems failing. A similar catalogue of seemingly divergent events can be made for the *Challenger* accident. In this case the web of events center on Halley's comet, President Reagan, a U.S. congressman, bad weather, Vice-President Bush, an erroneous weather forecast, a broken bolt, a flat battery, record cold temperatures, a sprinkler system, an O-ring, and a record wind shear.

The events of that fateful January morning when the *Challenger* was launched for the last time are probably familiar to most. Following is a brief account.

1986 was to have been an important year for the space shuttle program with a record fifteen flights scheduled, including the launch of a number of major scientific experiments (e.g., the ASTRO mission to study Halley's comet and the Hubble Space Telescope). The year got off to a bad start.

The launch of the shuttle *Columbia* on 12 January was the worst-delayed flight in space-shuttle history, with three schedule "slips" and four launch pad "scrubs." After this flight, *Columbia* was to be made ready for the launch of the ASTRO probe to observe Halley's comet. This mission had a firm deadline of 6 March, not only because this would provide best viewing of the comet, but also because it was three days before the much more ambitious unmanned Soviet mission would encounter Halley's comet. The delays in the *Columbia*'s January flight jeopardized the next scheduled flight, that of the *Challenger*, and, by way of a kind of domino effect, the subsequent relaunching of the *Columbia* for the Halley's comet mission. The *Challenger* could not take off until the *Columbia* had landed because spare parts from the *Columbia* were needed for the *Challenger*. The *Challenger* flight could not be canceled because on board was Teacher-in-Space Christa McAuliffe, who was planning to broadcast from space in time to coincide with President Reagan's State of the Union Message. The *Columbia*'s January flight could not be cancelled because on board was "pay-load specialist," U.S. Congressman Bill Nelson. Nelson was chairman of the House Space Science and Applications Subcommittee, which approves NASA's budget. The potential political damage to NASA and to Nelson (who had thirty-seven bus-loads of supporters at the Cape to watch the launch) of canceling the *Columbia* flight, even though it had gone through three time-consuming scrubs, was obvious. Although NASA tried to bring the *Columbia* home earlier, bad weather intervened and the flight had to stay up longer than expected. This led to two slips in the planned *Challenger* launch date.

The *Challenger* was finally set for launch on Sunday, 24 January, a date that would allow Vice-President Bush to watch the launch en route to the inauguration of the new

President of Honduras. However, Bush could not be kept waiting and needed to know in advance if a launch was in prospect. The weather forecast for Sunday was bad and it was decided on the Saturday afternoon to postpone the launch for a day and cancel Bush's visit. The weather forecast proved wrong and Sunday turned out to have been perfect weather for a launch. If there had not been the added pressure of Bush's visit NASA would have waited until the Sunday morning before deciding whether to go ahead with the launch. In that case *Challenger* would most likely have been launched that Sunday.

On Monday, 26 January the *Challenger* was finally tanked up with the astronauts strapped in and ready to go. Unfortunately a crucial bolt on the main hatch was found to have a sheared thread. Technicians called for a drill; a Black and Decker drill with a battery pack duly arrived. Unfortunately the battery was flat. More batteries were sent for. All were flat apart from one which was found to be too weak to drill the bolt. Finally an ordinary power drill was used (against strict safety regulations) but by this time the launch window had passed. The forecast for the next morning, Tuesday, 27 January, predicted low temperatures. The sprinkler systems were turned on the night before to help prevent ice building up in the troughs on the launch pad. The weather turned out to be much colder than anyone expected and all the water from the sprinkler system froze, making the icing problem even worse. Record low temperatures occurred in the lower part of the right-hand rocket booster which was shielded from the sun. The O-ring resilience in this area was severely impaired. At launch the O-rings gave way allowing gases from the solid fuel rocket to escape. A temporary seal formed which prevented further escape of gases. Ice from the launch gantry at the same time fell on the vehicle causing unknown damage. Thirty-seven seconds into the flight the *Challenger* encountered the most violent wind shear ever experienced on a space shuttle mission. The vibration broke the temporary seal in the solid-fuel rocket booster and fifty-eight seconds into the flight hot flame and gases fanned out like a flamethrower playing on the area where the strut of the booster was attached to the orbiter. At sixty-four seconds the external tank containing hydrogen fuel was ruptured. At seventy-two seconds the strut (possibly weakened by ice damage) gave way and the rotating solid-fuel rocket booster ruptured an oxygen tank. The rest is history.

The above account, like the one given by Perrow for the fictional job interview accident, displays a series of complex interacting events. Some events are strongly coupled, such as the failure of the O-ring and the eventual detachment of the booster. Other events are linked more subtly, such as the presence of a U.S. congressman on the *Columbia* which meant that that flight could not be canceled and led to the initial delay in the *Challenger*'s launch date. We also see that safety systems, in this case the sprinklers, interacted in an unexpected way to actually pose an additional safety risk through ice forming.

How Do We Decide on a Definitive Scenario?

Although Perrow's discussion of the job interview accident is instructive in terms of introducing us in a straightforward way to the typical system accident, it tends to obscure a question which needs to be asked before we can identify what type of accident has occurred. This is the question of how we know that the version of events given in the scenario is correct. This is a particularly pressing problem because the account of "what happened" is, of course, always a reconstructed account made *post facto*.

The difficulties involved in reconstructing what went on even in routine motor accidents are legion, with witnesses often producing different and contradictory versions. In the case of road accidents, we have ways of dealing with conflicting accounts. Some witnesses are more credible than others and this is established either by the police or in court. The police and forensic scientists can also produce independent "facts" which can serve to challenge and undermine purely impressionistic versions of what went on. However, in the hands of a skillful lawyer, this forensic evidence is defeatable as a number of recent cases have shown (Smith and Wynne 1989). Even presidential commissions cannot always satisfactorily resolve conflicting accounts. For instance, the Warren Commission, set up to investigate the assassination of John F. Kennedy, has come under attack for downplaying evidence of the existence of a second assassin.

The account of the *Challenger* accident summarized above is based upon Malcolm McConnell's book, **Challenger:** *"A Major Malfunction."* Some aspects of this account have been contested. For instance, the degree of pressure exerted on NASA to launch before Reagan's State of the Union message has been a subject of much speculation. The White House naturally denies any such pressure, but as McConnell and others have argued it is not necessary for the White House actually to contact NASA on the morning of the speech for the pressure to exist. It is a factor which must have been known to many of the relevant NASA personnel who made the decision to launch. The delicate social and political ties which comprise part of a complex technological system are probably some of the hardest aspects to reconstruct in terms of an accident scenario; for this reason accounts of accidents often leave them out. The Rogers Presidential Commission which investigated the *Challenger* accident had surprisingly little to say about such matters. For instance, the procedures whereby the major contracts for the space shuttle were awarded and the vested interests of some of these companies were not delved into in any great depth. Yet commentators such as McConnell see such factors as being integral to the causes of the disaster.

Much of the evidence being dealt with in the case of technological accidents is liable to be of a highly technical nature. Cameras, sensors, chart recorders, voice recorders, computer telemetry and so on provide evidence which might be held to be less ambiguous. This, however, is to assume that technical facts speak for themselves and escape a process of interpretation. One of the lessons of the new sociology of science is that even the most technical and seemingly "hard" areas of knowledge can be questioned and disputed by experts.[4] One of the main findings of the sociology of science of the last two decades has been the surprising degree to which technical experts can disagree—

not only over theories, but also over matters of fact.

In the case of the *Challenger*, the scenario outlined above could be disputed in the technical areas. For instance one expert, Ali Abutaha,[5] has put the cause of the accident down to a breach in the wall of the solid-fuel rocket booster. The breach may have been the result of damage to the external tank attachment ring on the solid rocket booster, caused by structural deflections during dynamic liftoff. What do we do with this version of events? Do we simply side with the Rogers Commission and say that this expert has got it wrong? What if the expert has technical evidence supporting his or her viewpoint? How are we as sociologists to judge such matters? Even versions of the accident which seem to have little credibility can present a problem. McConnell cites a group who blame the *Challenger* accident, and the whole series of other space misfortunes that beset America in 1986, on a new Russian gravity wave weapon. What do we make of this?[6] Are we happy to rule it out of court just because it seems incredible? It is especially difficult when the very accident or chain of events we wish to explain would have been seen by the majority of technical experts before the accident as incredible. Even if there is a majority of technical experts who discount a particular technical claim, can we trust the majority of experts? *Quis custodiet ipso custodes?* Certainly in the early days of nuclear power the experts would have had us believe that it posed little risk.

Given that an accident in a technical system will involve matters of fact and theory which may be disputed, how can a scenario of events be constructed which will provide the definitive account of what happened in order to start work in classifying the type of accident which has occurred? This question is not trivial. Future safety recommendations often depend crucially upon how the accident was classified. For instance, if the *Challenger* accident is seen as a result of NASA pork-barreling contracts then it follows that it is in the area of safeguarding fair and open competition for shuttle contracts where safety recommendations should be made. On the other hand, in the unlikely event that a Soviet gravity wave machine *was* being used, then presumably the safety recommendations would lie in the areas of elaborate shielding and effective countermeasures. Similarly, in the last analysis the policy recommendations which Perrow makes, such as the abandonment of nuclear power, follow from his ability to reconstruct accident scenarios.

Interpretative Flexibility

The problem encountered here is similar to the dilemma which faces a sociologist of science or technology studying a contemporary scientific or technological controversy. If there is one group of experts who argue in seemingly good faith and with a plethora of supporting facts and theories that, say, cold fusion exists and another group who argue conversely and in equal good faith that it does not, can the sociologist side with one side or the other? Close examination of these controversies reveals that the technical arguments alone rarely settle matters (Mulkay 1979; Collins 1985; Shapin and Schaffer 1985). At the heart of the social constructivist viewpoint is the notion of interpretative flexibility. This is the idea that scientific or technological facts can be given different meanings by different actors.

The *Challenger* accident provides an obvious example of how the meaning to be given a technological artifact can change dramatically:

> The average person viewing *Challenger* on Launch Pad 39B would have perceived a technological marvel, the quintessence of this country's innovative scientific excellence. In reality the vehicle—and especially its vulnerable and hazardous propulsion system—was the ultimate product of blatant Congressional pork barrelling, bureaucratic duplicity, inexcusable corporate deception, and public ignorance (McConnell 1988: 10).

This illustrates the notion of interpretative flexibility of technology in its most general sense. It is not the *Challenger* which has changed but people's perceptions of the technology. Indeed, the interpretative flexibility notion works through and through for the *Challenger* case. We find O-rings which exhibit interpretative flexibility. To some (e.g., NASA officials) they are fail-safe seals with redundant safety features, while to others (e.g., Morton Thiokol engineers) they are dangerously flawed. We find launch commit criteria which for NASA are the ultimate in safety criteria because they formally specify the conditions that must be met before the shuttle can be launched, yet for the Rogers Commission the constant waiving of these criteria (always for good reasons to the person doing the waiving) means they only give an illusion of safety and actually present an additional hazard.[7] Even recommendations made within the elaborate processes of consultation before going ahead with a launch can be given different interpretations. For example, Rockwell's statement on the morning of the fateful launch of the *Challenger* that the ice was such that they could not guarantee the safety of the vehicle was intended as a no-launch recommendation but was interpreted by NASA officials as a recommendation to launch.

The problem posed by the existence of interpretative flexibility is that participants can often give good reasons why they are interpreting things the way they do. This presents a problem to the sociologist studying such accidents. How can the sociologist provide an interpretation of events when the very events themselves are in dispute? The problem of assessing technical arguments becomes even more pressing when we start to enter the detailed technical arena of what happened to such and such a subsystem or component during the accident, and whether such and such a scenario or outcome could have been predicted and/or prevented.[8] Many such technical disputes were encountered by Perrow during his investigation into the Three Mile Island accident. Here are some examples:

> Apparently our knowledge of these plants is quite incomplete, for while some experts thought the closed valve constituted an important operator error, other experts held that it did not make much difference whether the valve was closed or not (Perrow 1984: 19).
>
> As an indication of how little we understand nuclear systems, I should note

> that shortly after the accident, some even urged that it was fortunate that the operator cut back on HPI [High Pressure Injection] although this was not the majority view (*Ibid:* 24).
>
> Contrary to industry pronouncements, there is still a good bit of mystery about atomic power plants, and this was an unfortunate case, since it was hours or days (depending upon whose testimony you wish to believe) before the bubble was conceived by the experts (*Ibid:* 29).
>
> To understand the accident, they would have to know that the core was seriously uncovered, and that a zirconium-water reaction was likely (a possibility disputed by an expert) (*Ibid:* 31).

Perrow treats these disputes as further evidence of the uncertainties surrounding nuclear power. In short, he treats experts disagreeing not so much as a problem for his analysis, but more as an indication that the technology is complex, uncertain, and inherently risky. The difficulty I find with this is that technical disagreement and expressions of uncertainty can be found surrounding virtually any technology which has been subject to an accident (Wynne 1988).

If it is the case that areas of technology are characterized by such technical disagreements and expressions of uncertainty then is there a way of dealing with them apart from bracketing off the whole area as being too much an object of contention to be amenable to sensible analysis?

Expertise and the Rogers Commission

Resolving these technical disputes is, of course, just what accident investigations set out to do. In the case of the Rogers Commission, its members included a number of technical experts. Attention has been drawn to the role and extent of this expertise:

> One important factor in the Commission's success was the expertise of its members, which included Neil Armstrong, a former astronaut, Sally Ride, a current astronaut, and Richard P. Feynman, a Nobel Prize–winning physicist. In fact twelve of the thirteen commissioners (all but Rogers) had technical expertise related to space travel in one form or another. . . . Their expertise prevented witnesses from misleading commission members. For example, when the project manager for the booster rocket, Lawrence Mulloy, justified his decision to authorize the launch, despite doubts about the O-rings, by claiming that "tests and analyses suggested that the seals could suffer three times the observed erosion and still block gases at far higher pressure than exists in the rocket" . . . the Commission was able to discount Mulloy's explanation, based on a technical evaluation of the relative quality of Mulloy's testimony about O-rings compared with that of the Thiokol engineers (Rowland 1986: 143).

Richard Feynman in his separate report on the *Challenger* accident makes the so-called mistaken reasoning employed by NASA officials more explicit:

> [T]here are several references to previous flights; the acceptance and success of the flights are taken as evidence of safety. But erosion and blow-by are not what the [O-ring] design expected. They are warnings that something is wrong. The equipment is not operating as expected . . . the fact that the danger did not lead to catastrophe before is no guarantee that it will not next time, unless it is completely understood. When playing Russian Roulette, the fact that the first shot got off safely is of little comfort for the next. . . . [O]fficials behaved as if they understood them, giving apparently logical arguments to each other—often citing the "success" of previous flights (Feynman 1989: 223).

To a NASA official such as Mulloy, these arguments were not "apparently logical," but perfectly valid reasons to discount the Thiokol engineers' worries. As Mulloy stated to the commission,

> My assessment at the time was, that we would have an effective simplex seal, based upon the engineering data that Thiokol had presented, and none of these engineering data seemed to change that basic rationale (Mulloy, quoted in *Report of the Presidential Commission* 1986: vol. 1, p. 92).

The argument the night of the famous tele-conference, as Thiokol engineers tried to convince the Marshall Space Flight Center that the O-rings would be compromised at low temperatures, centered upon the correlation between temperature and O-ring performance. The Thiokol engineers were forced to admit that they could not demonstrate any straightforward temperature dependence, and indeed, one of the worst cases of O-ring blow-by had occurred at high temperatures. Mulloy and his boss at Marshall, George Lucas, were famous for basing their decisions only upon quantitative data.

If we follow the intricacies of the debate between the people at Thiokol and those at Marshall, it is very hard to say that Mulloy was in any obvious sense wrong. It turns out that the Marshall group had produced a mathematical model to fit the erosion data and that this suggested a safety factor of three. In his report Feynman pulls this model apart demonstrating the various uncertainties of the parameters involved. However, it is doubtful whether Feynman's is the last word in the argument. The trustworthiness of mathematical models is an issue to occur in other areas of science (Pinch 1986); Feynman's claim that we should not trust mathematical models seems to belie the fact that often scientists do rely on such models.

I am not saying that Mulloy *did* have adequate grounds but rather to point to the problems we as social scientists face in assessing such arguments. It would seem to be a very tame sociology that merely accepted the technical recommendations made by such accident investigations.

We should also note that it is naive in the extreme to think that it is technical arguments alone which settle such matters (see Gieryn and Figert 1990). Feynman is well known as a master rhetorician—his use of the analogy of Russian Roulette is very powerful but is it strictly a technical argument? If we read Feynman's account of his involvement in the Rogers Commission published in his posthumous book, we find a typical Feynman-esque combination of blather and brilliance. Feynman compares himself with other well-known physicists who are famous for shooting-down "crazy ideas," such as R. W. Wood, who played a large part in the demise of N-Rays. It was in thinking of such cases that Feynman claims he got his idea for his famous experiment where he demonstrated to the world the effect of iced water on O-rings. This experiment was, of course, extraordinarily effective as a piece of public rhetoric, but it is doubtful whether it added anything by way of a serious technical contribution. Here is Feynman's honest-joker account of what he did:

> So finally, when I get my ice water, I don't drink it! I squeeze the rubber in the C-clamp, and put them in the glass of ice water. After a few minutes, I'm ready to show the results of my little experiment. I reach for the little button that activates my microphone. General Kutyna [a fellow commissioner], who's caught on to what I'm doing, quickly leans over to me and says, "Co-pilot to pilot: not now." Pretty soon, I'm reaching for my microphone again. "Not now!" He points in our briefing book—with all the charts and slides that Mr. Mulloy is going through—and says, "When he comes to this slide, here, that's the right time to do it." Finally Mr. Mulloy comes to the place, I press the button for my microphone, and I say, "I took this rubber from the model and put it in a clamp in ice water for a while" (Feynman 1989: 151).

The effect of Feynman's demonstration is well known (Gieryn and Figert 1990). Feynman became the instant darling of the media and there can be little doubt that his argument was effective. NASA was made to look as if they had foolishly overlooked an obvious piece of evidence. However, it is far from clear that it was simply technical expertise that carried the day.[9]

Thus far, I have suggested that the debate over the merits of Mulloy's arguments shares characteristics with other technical disputes and I have pointed to the difficulty the sociologist faces in assessing such debates. Yet the sorts of accidents we are dealing with and the policy implications which follow depend upon the outcome of such technical disputes.

Sociology of Science and the *Challenger* Accident

Interestingly some sociologists of science have commented on the *Challenger* accident, and perhaps even more interestingly, their conclusions can be taken to support the reasoning employed by both Mulloy and the Thiokol engineers with whom Mulloy was in dispute.

It is commonplace within the sociology of science to draw attention to the uncertainties in scientific knowledge experienced by the practitioners who are most familiar with that knowledge. This argument has been made most strongly by Harry Collins:

> Certainty about natural phenomena tends to vary inversely with proximity to scientific work . . . proximity to experimental work . . . makes visible the skillful, inexplicable and therefore potentially fallible aspects of experimentation (Collins, quoted in MacKenzie et al. 1988).

Thiokol engineers, such as Roger Boisjoly, have gone on record as saying that their engineering "feelings" that the O-rings were a problem could not be quantified. This could be taken as support for Collins's claim that those closest to the experimental work and most familiar with the O-rings are most likely to be aware of the uncertainties that they presented. Donald MacKenzie and colleagues, quoting Collins, have made this point explicitly in relation to the *Challenger* accident:

> A worrying analogy is the decision to launch the space shuttle *Challenger* on its fatal mission, for it was precisely from the "certainty trough" [i.e. those people who did not have hands-on experience of the O-rings] that the decision makers were drawn. They had a deep commitment to the technological institution involved, but were insulated from the uncertainties of those with direct responsibility for producing knowledge about the safety of the solid booster rocket (MacKenzie et al. 1988: 162).

This argument can be taken to support Boisjoly. In other words Mulloy, the manager of the solid-booster program, was not the person getting his hands dirty actually studying the O-rings and having to face the uncertainties they presented. His distance from the scientists actually looking at the O-rings meant that he placed more confidence in their reliability than the Thiokol engineers.

However, Brian Wynne in a recent article takes a rather different tack. He argues that it is always the case that engineering practices do not follow formally prescribed rules. As he puts it,

> In nearly every investigation of accidents and their precursors, or in good ethnographic research within normally operating technological systems, one finds the same situation. Beneath a public image of rule-following behaviour, and the associated belief that accidents are due to deviation from these clear rules, experts are operating with far greater levels of ambiguity, needing to make uncertain judgments in less than clearly structured situations. . . . Practices do not follow rules; rather, rules follow evolving practices (Wynne 1988: 153).

The logic of Wynne's argument is that most technologies operate in a much more ad hoc way than we realize and that, rather than having a reaction of shock and horror that the official standards have been breached (the typical reaction to the shuttle accident), we should realize that this is the nature of the beast. He writes:

> [I]t was already known by practitioners that several previous shuttle flights had shown signs of these O-ring faults. . . . Empirical experience was that this abnormality had happened in several previous tests and real flights, without leading to an accident. On the fatal launch, therefore, the logic of the decision was not that the O-rings had shown signs of failure before, and performed at less than their official performance–rule, and therefore that all further launches must be stopped until this subsystem was redesigned. Rather the logic was: this component shows behaviour which is abnormal according to our original design–performance rules; however, in several launches it has shown less than adequate performance without incident; its failure has apparently been within acceptable bounds (which we have made up under negotiation with experience as we went along) (Wynne 1988: 151).

Thus, according to Wynne, Mulloy was doing no more than was normal in the management of an unruly technology. It was the commission and Feynman who had taken up the official discourse of technology, the discourse of formal rules and checks and absolute certainty, with which they castigated Mulloy and NASA.

To add emphasis to Wynne's argument, he points out that the O-rings were just one subsystem or component among many others on the shuttle that were offering similarly ambiguous decision-judgment problems. If the formal procedures were followed all the time the technology would soon grind to a halt—certainty just cannot be guaranteed.

The different conclusions which I have drawn from Wynne and MacKenzie (and here I am probably pushing their analysis further than either of them would want to see it taken) seem to follow paradoxically from the same types of analytical moves as are made by Perrow. Rather than relying upon organizational analysis we now have sociology of science to tell us which areas of knowledge are more certain than others. MacKenzie is more certain about Boisjoly's uncertainty claim regarding the O-rings because of the sociology of science discovery that uncertainty is to be found among those who practice hands-on research. Wynne is more certain that Mulloy's rationale for proceeding to launch is justified because this is the way that unruly technologies have to be operated in practice.

That MacKenzie and Wynne place certainty in different areas is itself a reminder of the fundamental difficulty sociologists face in assessing these uncertainty arguments. Within science rhetorics of certainty and uncertainty can be used for a variety of purposes (Pinch 1981). The dilemma for the sociologist is that when certainty and uncertainty become themselves rhetorical tools to be used in the debate how can we draw any firm conclusions or recommendations as to safety?

A Final Note on the Space Shuttle Case

The problems involved in assessing the sorts of arguments I have been discussing in this paper have not gone away. Today there are allegations that with the implementation of the findings of the Rogers Commission an enormous and unwieldy safety bureaucracy has been created within NASA which will have a paralyzing effect and prevent the organization from operating an effective and efficient space shuttle program. Also, the difficulties of assessing expert arguments and claims of uncertainty remain. Roger Boisjoly, the whistleblowing Morton Thiokol engineer, has been placed on indefinite disability leave and is involved in a billion-dollar personal injury and damages suit against Morton Thiokol. According to Boisjoly, the redesigned O-ring joint still presents an unacceptable safety hazard. The one hundred high-strength bolts placed around the surface of the joint as the ultimate safety redundancy are, according to Boisjoly, a possible source of unpredictable stress and the new bolts represent possible fire-paths. Boisjoly is quoted as saying

> Murphy's Law awaits them. If they're right then they are heroes and I'm a bum. And that's okay. If I'm right, more people could die (Boisjoly, quoted in McConnell 1988: 262).

The certainty we should attach to Boisjoly's claim of uncertainty seems to me to be as difficult to assess as ever.

Conclusion and Caveats

Since large-scale systems are integrally social and technical in character, it seems to me that sociologists must meet the challenge of dealing with technical uncertainty. The difficulty I have with Perrow's treatment of the issue is that it tends to avoid biting the bullet of technical uncertainty by going for carte blanche recommendations in areas where such uncertainty is encountered. This may be the prudent course of action but my question for Perrow would be that if the weakly coupled and more desirable nuclear power plants he refers to (Perrow 1984: 38) are developed and we get technical experts disagreeing as to their performance, how will he be able to evaluate such arguments?

It can be seen in this paper that as we have gone further and further into the technical arguments a paradox has arisen. The knowledge of my own field—sociology of science—has itself come into question, with conflicting recommendations being derived from it. What does this fundamental uncertainty in the application of the sociology of scientific knowledge to the *Challenger* case imply? It shows, of course, that our own findings are themselves subject to interpretative flexibility—a finding which thereby confirms this very analysis![10]

This is hardly a surprise, since it would be strange if we ended up saying that space shuttles may or may not work, that nuclear power has different degrees of risk attached to it depending upon which expert you talk with, but that sociology is the one certainty we have.

Readers of this paper may be perplexed by this conclusion. I do not claim to have shown an adequate solution to the problem of technical uncertainty, but I do hope to have shown that it is a salient issue for anyone concerned to understand large technical systems and their development.

NOTES

1. I have had many helpful discussions on the *Challenger* case with Michael Mulkay, Malcolm Asmore, and Tom Gieryn. I would also like to acknowledge useful discussion of this paper on the occasion of its presentation at the Large Scale Technological Systems Conference, Berkeley, California, 17–21 October, 1989.

2. Other shuttle flights have carried nuclear-powered satellites and if a similar accident had happened to them the consequences could have been much more dire.

3. One survey of the American populace reported the accident as having a greater impact in its aftermath than a number of recent disasters where loss of life was more severe, e.g., the Beirut barracks bombing.

4. For example, in 1968, a supposedly reliable experiment set up to measure one of the most well-understood properties of the sun—its production of neutrinos—failed to find the expected fluxes of neutrinos. Consequently over four hundred scientific papers have been published offering different accounts of the cause of this "accident." Even areas of physics previously held to be well understood and unproblematic have come under challenge (Pinch 1986).

5. It is reported in *Flight International,* 18 September 1987, p. 16, that "AbuTaha's studies have been published widely, and his theories have gained much support in some aerospace circles. . . . NASA, however, is adamant that the Rogers Commission got it right."

6. For an approach which takes even the incredible seriously, see our account of the reception of Uri Geller among the physics community (Collins and Pinch 1982). The interesting thing about doing sociological fieldwork among people that we routinely write off as cranks is how sensible they seem when you talk to them!

7. For a provisional analysis of the interpretative flexibility in the O-rings and the Launch Commit Criteria see Pinch, Mulkay, and Ashmore 1986.

8. This issue was dramatically raised by the 17 October 1989 Loma Prieta earthquake, which occurred during our conference. On the local television an expert (in this case a graduate engineer) was interviewed who claimed to have carried out a study which showed that the Nimitz Freeway was liable to just the sort of earthquake damage which led to its collapse. Other experts had overlooked and/or presumably had reasons to ignore his findings.

9. For an account of the uses to which scientific experiment can be put see Gooding, Schaffer, and Pinch 1989; for further description and analysis of Feynman's O-ring experiment and how it worked as a piece of scientific rhetoric, see Gieryn and Figert forthcoming.

10. For an attempt to take this self-referential quality of the sociology of science into the field of policy see Ashmore, Mulkay, and Pinch 1989. See also Ashmore 1989.

REFERENCES

Ashmore, M. 1989. *The Reflexive Thesis: Writing the Sociology of Scientific Knowledge.* Chicago: Chicago University Press.

Collins, H. M. 1985. *Changing Order.* London and Beverly Hills: Sage.

Collins, H., and Pinch, T. J. 1982. *Frames of Meaning: The Social Construction of Extraordinary Science.* London: Routledge and Kegan Paul.

Feynman, R. P. 1989. *What Do You Care What Other People Think?* London: Unwin Hyman.

Gieryn, T. F., and Figert, A. E. 1990. "Ingredients for a Theory of Science in Society: O-Rings, Ice Water, C-Clamp, Richard Feynman and the *New York Times.*" In *Theories of Science in Society*, ed. S. Cozzens and T. Gieryn. Bloomington: Indiana University Press.

Gooding, D.; Pinch, T. J.; and Schaffer, S. 1989. *The Uses of Experiment.* Cambridge: Cambridge University Press.

Lewis, R. S. 1988. ***Challenger:*** *The Final Voyage.* New York: Columbia University Press.

McConnell, M. 1988. ***Challenger:*** *"A Major Malfunction."* London: Unwin.

MacKenzie, D.; Rudig, W.; and Spinardi, G. 1988. "Social Research on Technology and the Policy Agenda: An Example from the Strategic Arms Race." In *Technology and Social Process, ed. B. Elliott.* Edinburgh: Edinburgh University Press, pp. 152–80.

Mulkay, M. 1979. *Science and the Sociology of Knowledge.* London: George Allen and Unwin.

Perrow, C. 1984. *Normal Accidents: Living with High Risk Technologies.* New York: Basic.

Pinch, T. J. 1981. "The Sun-Set: On the Presentation of Certainty in Scientific Life." *Social Studies of Science* 11: 131–58.

————. 1986. *Confronting Nature.* Dordrecht: Reidel.

Pinch, T. J.; Mulkay, M.; and Ashmore, M. 1986. "The Sociology of Machine-Based and Social Technologies: Bicycles, Space Shuttles, Penitentiaries, and Hospital Clinical Budgeting." Paper presented at the annual meeting of the Society for Social Studies of Science, Pittsburgh, Pennsylvania, October.

Pinch, T. J., and Bijker, W. 1984. "The Social Construction of Facts and Artefacts." *Social Studies of Science* 14: 199–241.

Report of the Presidential Commission on the Space Shuttle Challenger *Accident.* 1986. Washington, D.C. Five volumes.

Rowland, R. 1986. "The Relationship between the Public and the Technical Spheres of Argument: A Case Study of the *Challenger VII* Disaster." *Central States Speech Journal* 37: 136–46.

Smith, R., and Wynne, B., eds. 1989. *Expert Evidence: Interpreting Science in the Law.* London: Routledge.

Trento, J. T. 1987. *Prescription for Disaster*. New York: Crown.

Turner, Barry. 1978. *Man-Made Disasters*. London: Taylor and Francis.

Wynne, B. 1988. "Unruly Technology: Practical Rules, Impractical Discourse and Public Understanding." *Social Studies of Science* 18: 147–68.

NOTES TOWARD A SOCIOLOGY OF SUPERCOMPUTING

Donald MacKenzie
Department of Sociology
University of Edinburgh

Introduction

The tentativeness of my title is wholly deliberate. This paper is a report on work in progress. Though some features of the development of supercomputing already emerge clearly from the research, others take the form of tantalizing glimpses of possible patterns.

After an initial discussion of the meaning of "supercomputer," and of the main uses of such devices, I give a brief history of supercomputer development. There have been two main strands within supercomputing: a mainstream, characterized by a "reverse salients" pattern of technical change, akin to that identified by Thomas P. Hughes in his study of the evolution of electricity supply systems (Hughes 1983); and a set of radical alternatives. The paper then turns to three themes from the social studies of science and technology, asking what light they throw on supercomputer development and, reciprocally, what we might be able to learn about them by studying the history of supercomputing. The first theme concerns the concept of "technological trajectories"—an influential attempt to specify predictable patterns in technological development. I argue that the metaphor of "trajectory" misconstrues these patterns. It fails to take into account the centrality to technological development of issues of technical belief systems and self-fulfilling prophecies.

The second theme concerns the capacity of powerful social interests to shape technological development. The particular interest in question is that of the U.S. nuclear weapons laboratories, which were originally the prime users of supercomputers. Though their effect on the development of supercomputing has been considerable, the limits of their influence can also been seen.

The paper ends with speculative exploration of a third theme. All large technical systems raise the issue of control: how are they to be designed so that the action of their parts is coordinated? Intriguing parallels exist between the different solutions to this proposed for computer systems and solutions to the analogous problem of order in society. Though the importance of the parallels should not be exaggerated, there are hints in the history of supercomputing that this phenomenon is of importance in understanding the shaping of technical systems.

The research reported on was funded by the United Kingdom Economic and Social Research Council's Programme on Information and Communication Technologies.

T. R. La Porte (ed.), Social Responses to Large Technical Systems, 159–175.

What Is a Supercomputer?

At one level the definition of "supercomputer" is obvious. Supercomputers are the most powerful computers, at least at any given point in time. But that definition begs two important questions: how is "power" defined, and what is it used for?

The first question has a straightforward answer. The power of supercomputers is speed at floating-point arithmetic operations.[1] Though in the 1950s this was typically expressed as the time taken to perform a single such operation—most usually a multiplication. Now it is expressed as the number of floating-point operations carried out per second.

The performance level required to constitute a supercomputer has grown from the thousands (KFLOPs or kiloflops) in the 1950s, to millions (MFLOPs or megaflops) in the 1960s, thousand millions in the 1980s (GFLOPs or gigaflops) and, possibly, million millions in the 1990s (TFLOPs or teraflops).[2] Machines rated at sixteen to twenty gigaflops have been promised for release during the next twelve months. Several teams are developing machines they hope will by around 1993 get within at least an order of magnitude of the teraflop goal.

Most personal computer have speeds in the kiloflop range, and large commercial mainframe computers—of the type sold by IBM—are typically rated low in the range of megaflops. While there are many problems with both sorts of device, lack of arithmetic speed is not commonly listed high among them. So why are gigaflops, or even teraflops, of interest? What is the use of all this power? At its simplest, it is to simulate events and processes that are too big, too small, too violent, too inaccessible, or too expensive to find out about by other means. Examples include nuclear explosions, the weather patterns in the earth's atmosphere, oil fields, the aerodynamics of a yet-to-be-built bomber, and the world economy.

In a more general sense, supercomputing should be seen as a technology of power, as a resource for the domination of phenomena that are otherwise hard to subject to control—a resource that is often seen as central to the military (and increasingly economic) strength of states. It is interesting that the U.S. government treats supercomputing as a technology of power, using it, in the words of one Control Data executive, as "the carrot or the stick in the U.S. government's effort to reward or punish other governments in the realm of foreign policy" (Donaghue 1987: 188).[3]

The Development of Supercomputing

Supercomputing began in the latter half of the 1950s with the Univac LARC (Livermore Atomic Research Computer, or in some versions Livermore Automatic Research Computer) and the IBM Stretch, both machines designed to far surpass the existing state of the art in floating-point arithmetic speed.

The latter partially failed, and since then IBM has had an ambivalent attitude toward supercomputing, not wishing to see it pass entirely into the hands of competitors, but also being unwilling, at least until very recently, to launch ambitious new supercomputer projects. Accordingly, U.S. and world supercomputer leadership passed to the Minneapolis-based Univac spin-off, Control Data, and to its chief supercomputer designer, Seymour Cray. The 1964 Control Data 6600 and 1969 Control Data 7600 were the result.

Thereafter, mainstream supercomputing bifurcated. Control Data, though not Cray himself, took on an ambitious new supercomputer project, the STAR-100, designed to be capable of 100 megaflops. Cray's Control Data 8600 was stillborn, and Cray left Control Data to set up his own firm, Cray Research, designing the CRAY-1, CRAY-2 and the shortly to be released CRAY-3. Control Data itself went on to produce the CYBER 205 and ETA[10] supercomputers, before losses totalling $490 million forced the closure of its supercomputer business in April 1989.

Cray Research split in May 1989, after a decade of internal bifurcation, through which two competing lines of technical development had been pursued simultaneously—the CRAY-1, CRAY-2, and CRAY-3 series; and the Cray X-MP, Y-MP, and C-90, a line initially led by "number two designer" Steve Chen. The Control Data closure and Cray split became a matter of widespread public comment in the United States in the summer of 1989, because it suggested that the United States was faltering in the face of foreign competition. By then mainstream supercomputer development was well established in Japan, with three major firms—Fujitsu, NEC, and Hitachi—active in the area. The NEC SX-X, announced for summer 1990 delivery, will have a claimed top speed of 20 gigaflops—in excess of the CRAY-3, also slated for delivery in that period.

In these developments speed has been sought by two means. The first has been improved component technology reducing gate delay time, i.e., reducing "the time taken for a signal to travel from the input of one logic gate to the input of the next logic gate" (Hockney and Jesshope 1988: 3n). The "first generation" electronic valve computers of the early 1950s had gate delays of around a microsecond; the fastest integrated circuits of the mid-1970s permitted that to be reduced to around a nanosecond.

That three-orders-of-magnitude improvement cannot, however, on its own account for the roughly five orders of magnitude increase in processing speed over the period (Hockney and Jesshope 1988, p. 4). The other two orders of magnitude can be attributed to changes in computer architecture—the gradual introduction of various forms of parallelism or concurrency. Six of these deserve special mention: concurrent input/output operations; pipelining; memory interleaving and hierarchy; parallel functional units; vector processing; and multiple central processor units.

Providing specialized hardware and software so that input of data and programs and output of results can go on concurrently with processing both predates and is more widespread than supercomputing. In the search to eliminate all barriers to speed, it was nevertheless developed to a considerable degree in supercomputing.

Pipelining, introduced with the earliest of the supercomputers such as Stretch, was originally specific to supercomputing. In a non-pipelined computer the different phases of the execution of a single instruction—accessing and interpreting the instruction, accessing the operands, performing the operation, returning the result to memory—are performed one after the other. In a pipelined computer they are overlapped, so that while one instruction is being interpreted, another is being accessed, and so on.

Memory interleaving and hierarchy—also early and widespread techniques—are designed to prevent accessing instructions or data from memory, or writing results to memory, becoming a bottleneck. In interleaving, memory is arranged so as to allow simultaneous access to different segments of memory. In memory hierarchy, small amounts of ultrafast (and expensive) memory are provided in addition to the slower (and cheaper) main memory. This is to enable as many transfers as possible between memory and central processor to involve only the small, fast memory, not the main memory.

Separate specialized units for addition, multiplication, division, etc., that can operate independently and in parallel, was a particular feature of the Control Data 6600, which contained ten parallel functional units: a Boolean unit, a shift unit, a fixed-point adder, a floating-point adder, two multiply units, a divide unit, two increment units, and a branch unit.

Vector processing is the simultaneous execution of a single instruction on all the members of an ordered set of data items. The first proposal for a pipelined vector computer was the Control Data STAR-100. Though conceived in the mid-1960s, it was not operational until 1973 (Hockney and Jesshope 1988: 21). The first unequivocal success in pipelined vector computing, however, was the 1976 CRAY-1.

The most recent parallelism development—constructing supercomputers with more than one central processing unit—was introduced in the 1980s. To date, two, four, and eight units have been the most common choices. In the near future, we are very likely to see significantly larger numbers of units configured into a single supercomputer. While this is a potentially major step in parallelism, in practice these installations often seem to be run primarily as a collection of separate computers. Rather than the components of a single task being distributed over all processors, each processor individually runs unrelated tasks, such as different programs for different users.

With the partial exception of vector processing and multiple central processing units, these forms of parallelism represent piecemeal, incremental alteration of the fundamental sequential computer architecture that has become associated with the name of John von Neumann. Cray himself put matters succinctly in 1970 (in his foreword to Thornton 1970). His Control Data 6600, he said, attempted to "explore parallelism in electrical structure without abandoning the serial structure of the computer programs. Yet to be explored are parallel machines with wholly new programming philosophies in which serial execution of a single program is abandoned." Even vector processors or multiple central processing units, while allowing a considerably greater degree of parallelism in program execution, did not wholly abandon this.

The development of supercomputing bears a considerable resemblance to the patterns found in the development of electricity supply networks and a range of other technologies by Thomas P. Hughes (see especially Hughes 1983). This pattern of change involves pursuit of a single dominant objective: growth in the case of electricity supply networks, floating-point arithmetic speed in the case of supercomputers. At each stage of development the predominant barriers to progress towards the goal—"reverse salients" in Hughes's terms—are sought, and innovation focuses on removing them.

For example, Cray's Control Data 6600 and 7600, despite their pipelining and multiplicity of parallel functional units, could not perform floating-point arithmetic faster than at a rate of one instruction per clock period.[4] This "operation issue bottleneck . . . [was] overcome in the CRAY-1 processor by the use of vector orders, which cause streams of up to 64 data elements to be processed as a result of one instruction issue" (Ibbett and Topham 1989: vol. 1, p. 113).

"Mainstream supercomputing," exemplified by Control Data, Cray, and the Japanese companies, has certainly taken substantial and daring steps—vector processing being perhaps the clearest example—but nevertheless remains a fundamentally incremental form of technological change. Substantial increases in speed are gained by gradual modification of existing computer architectures—in response to perceived bottlenecks or reverse salients—rather than by scrapping what exists and adopting a radically different architecture.[5]

The latter path, however, has also had its proponents. While they do not agree on the best alternative architecture, their machines have shared a common feature—the use of more and different kinds of parallelism than in "mainstream supercomputing."[6] The most important example is the processor array. The key figure in the development of processor arrays was Daniel L. Slotnick. Slotnick's idea was to build a computer that would perform the same operation or sequence of operations concurrently on many units of data. His first system was SOLOMON, "because of the connotation both of King Solomon's wiseness and his 1000 wives" (Slotnik 1982: 22). It was to have 1,000 separate processing elements, each performing the same fixed-point arithmetic operation concurrently on different data (Slotnick, Borck, and McReynolds 1963); in the later terminology of Flynn (1972) it was to be a single instruction stream–multiple data stream (SIMD) parallel computer.

The SOLOMON design evolved into an even more ambitious scheme, ILLIAC IV (Barnes et al. 1964; Slotnick 1971). The number of processing elements shrunk to 256, arranged in four quadrants of sixty-four, but they were no longer the original simple fixed-point processor. Each would be capable of concurrent operation on all the bits of a 64-bit floating point number. The overall performance goal, unprecedented at the time, was a gigaflop, and "Illiac IV ultimately included more than a million logic gates—by far the biggest assemblage of hardware ever in a single machine" (Falk 1976: 65).

The failure of suppliers to produce the required integrated circuits, anti-war demonstrations, riots and firebombing on the campus of the University of Illinois in 1970, and a range of other circumstances prevented the full development of ILLIAC IV (Falk 1976; Slotnick 1982). Eventually only one 64-element quadrant was installed at the NASA Ames Research Center in California in 1972, where it became a centerpiece of the first modern worldwide computer network, the ARPA Net.[7]

ILLIAC was the harbinger of a wave, a decade later, of highly parallel challengers to mainstream supercomputing, a wave that included MIMD (multiple instruction stream–multiple data stream) machines as well as SIMD ones. This wave was reinforced by the technology of integrated circuits, the fundamental component of the last two decades' computers, based on the better-known of its two forms of development.

This better-known form is founded upon field-effect transistors, in which current flows only in the surface plane of the microchip. The most common are MOS (metal-oxide-silicon) chips, fabricated by depositing first a layer of oxide and then a layer of metal on a silicon semiconductor substrate. Their original uses were in simple devices such as calculators and digital watches (Shankar and Fernandez 1989: 17). Field-effect technology offers relatively easy mass production manufacture. As production techniques such as photolithography have steadily improved, many more components can be fabricated on the one chip.

The great drawback of field-effect technology from the point of view of supercomputer designers has been its relative slowness, or lengthy gate delay.[8] Cray originally eschewed integrated circuit technology altogether. When he did move to the integrated circuit in the 1970s, it was not to field-effect technologies such as MOS, but to the alternative form of microchip, bipolar technology.

Bipolar chips were the original integrated circuits used in computers. In these chips, electric current flows perpendicular to the surface of the chip, as well as along it (as in a field-effect chip). Different regions of a bipolar chip are "doped" with different impurities. Bipolar chips are harder to fabricate than field-effect chips, and much smaller levels of integration are possible, perhaps two orders of magnitude less. But, crucially for supercomputer manufacturers, the gate delays in bipolar chips have been considerably less than in field-effect chips. Hence mainstream supercomputing adopted bipolar technology, particularly emitter-coupler logic. With his CRAY-3, Seymour Cray is about to make a further move in this direction with the adoption of the even faster, but even harder to fabricate and more expensive, technology of gallium arsenide.

However, the wider dominance of the less demanding, more highly integrated MOS technology opened up an opportunity that during the 1980s was seized by literally dozens of projects. By 1986, for example, MOS chips constituted about 78 percent of semiconductor production, with the once dominant bipolar technology squeezed to a mere 12.5 percent, and gallium arsenide still negligible (Shankar and Fernandez 1989: 17). Why not adopt MOS, benefit from its ease of fabrication and economies of scale, and try to compensate for the relative slowness of individual MOS chips by configuring large numbers of them in highly parallel architectures? As one important early paper put it,

> The premise is that current LSI [large-scale integration] technology would allow a computational facility to be built around a large-scale array of microprocessors. . . . We anticipate that individual microprocessors would use a technology with intermediate values of gate speed and gate density to keep costs low. Therefore, the individual microprocessor circuitry is likely to be of only moderate speed. Total processing speed and throughput for the entire system would be obtained through parallelism. The network itself might contain as many as 2^{14} (= 16,384) microprocessors to obtain a very high degree of parallelism (Pease 1977: 458).

The result in the 1980s was an explosion of parallel architectures, such as the various "hypercubes" (discussed below) and the Connection Machine, or the machines built around the

UK Inmos transputer, a very-large-scale integrated complementary metal-oxide semiconductor (CMOS) chip designed specifically for parallel processing (Johnson and Durham 1986; Molina 1989).

Until very recently, these projects did not seek to rival mainstream supercomputing in absolute floating-point performance, but rather to offer a superior price-performance ratio. But with the most advanced field-effect chips, such as the one-million transistor Intel i860 now being claimed to offer on a single chip the processing power of a 1976 CRAY-1, rivalry in absolute performance is growing. Thus the "Touchstone" project announced in April 1989 aims at the parallel configuration of up to 2,048 such chips (Intel Scientific Computers 1989). In June the Oregon firm NCube announced what it claimed to be the world's fastest computer, the NCube 2 Scalar Supercomputer, with up to 8,192 processors connected together.

Natural Trajectories of Technology?

The question of supercomputer components raises the first general issue I want to discuss, that of technological trajectories. Supercomputer designers are often unable or unwilling to consider merely organizing and interconnecting components that are already in existence. These components are expected to improve rapidly. It may well be an advantage to design for components that will become available in the future. If the timing is right, it gives one's machine a competitive edge; on the other hand, if the desired components do not become available, disastrous delay may result.

The supercomputer designer has, therefore, often to be in the business of prediction. What transistors or integrated circuits will be available in, say, two years' time? What balance is there between the risk of being overtaken by a competitor using more advanced components, and the risk of a project failing through the components for it not being ready?

The point is of general interest because, in the United Kingdom at any rate, there has been a considerable vogue in the analysis of technical change in terms of "technological trajectories." The vogue is largely among economists, who have been reacting against the notion that technical change is plastic, mere putty to be molded this way or that by market forces. There are, says one such writer, "directions of technical development that are cumulative and self-generating" (Pavitt 1984: 355).

It is hard to see a better candidate for "technological trajectory" than the development of computer components.

> In 1964, noting that since the production of the planar transistor in 1959 the number of elements in advanced integrated circuits had been doubling every year, Gordon E. Moore, who was then director of research at Fairchild, was the first to predict the future progress of the integrated circuit. He suggested that its complexity would continue to double every year. Today, with circuits containing 2^{18} (= 262,144) elements available, we have not yet seen any

significant departure from Moore's law (Noyce 1980: 33–34).

Computer designers do indeed seem to employ notions such as this (albeit much more detailed and sophisticated) in making their crucial predictions about what components will be available to them to organize and interconnect. To that extent, then, they believe in technological trajectories. However, there is an implication of this which has not received proper attention in the literature on technological trajectories: the beliefs people like computer designers hold in the future path of technical development help to shape that technical development.

If people make their plans around components of a particular type becoming available, place advance orders for them and support their development in other ways, then they help provide the conditions that bring the predicted technology into being. This, for example, is currently the case with the first supercomputer logic circuits fabricated out of gallium arsenide (Cray 1988).

At the core of a technological trajectory is thus a successfully self-fulfilling prophecy. The crucial factors turning the potential anarchy of technological change into a persistent trajectory are the expectations and interests of potential users, who build into their calculations the belief that technological change will take a particular form and then invest money, careers, and credibility in the predicted form of change, thus helping it to happen. Finding their predictions correct, they continue to act on them with confidence and encourage further "self-prophesied" progress.

In noting the presence of a self-fulfilling prophecy in a technological trajectory, I do not intend to argue that technological change is wholly plastic. Configurations of social interests, however powerful, will not always be able to give technology a desired shape. Some extremely widely held predictions turn out to be false. The material world is obdurate.

Yet there is an important phenomenon that is glossed over when technological trajectories are thought of as somehow "natural" (see Nelson and Winter 1982: 258–62). The sense that a path of technical development is natural, that it corresponds to the inherent possibilities of a technology, is a post hoc effect of the self-fulfilling prophecy. Those lines of technical development that are not pursued do not improve; those that are pursued often do.[9] The progress of the latter, and lack of progress of the former, can then appear to be due to the inherent possibilities of a technology.

This analysis of technological trajectories as self-fulfilling prophecies is a sociological one in two senses. Previous sociological theorizing often took the self-fulfilling prophecy as a pathological phenomenon (see the classic discussion in Merton 1948). Recent work, particularly Barnes 1983, has argued that it is pervasive and normal. Any stable social institution—money is a good example—is a self-fulfilling prophecy. This way of thinking about technological trajectories thus helps us understand the processes of the institutionalization of particular forms of technical change.

In addition, it points to the importance of understanding technical belief systems and credibility. What persuades technologists that some things will happen and that others will not? Whose predictions are believed and why? To the extent that these phenomena have been investigated at all, it has typically been for exceptionally controversial technologies, and often the beliefs of the public, rather than the technologists themselves, have been the focus. We need a deeper understanding of how these belief systems operate under normal circumstances within technological communities.

Nuclear Weapons Laboratories and Supercomputer Architectures

One "social interest"—nuclear weapons development—more than any other was involved in the early evolution of supercomputing. The first supercomputers, LARC and Stretch, were developed in close collaboration with the two U.S. nuclear weapons design laboratories at Livermore and Los Alamos. LARC was commissioned by Livermore. Stretch was proposed by IBM to Los Alamos, and its staff took part in drawing up its specifications and technical design. Four of the eight Stretch computers sold went to organizations heavily involved in nuclear weapons research. Three others supported military work, and one was sold to a U.S. weather bureau (Bashe et al. 1986: 456).

The nuclear weapons laboratories remained vitally important as customers. Los Alamos and Livermore have bought about ten of their preferred machine from each generation of supercomputer—the Control Data 6600, 7600, CRAY-1, and Cray X-MP. For the earlier machines this was between about 10 percent and a quarter of total sales. Only with the X-MP, the most widely sold Western supercomputer to date, have sales to the weapons laboratories formed significantly less than 10 percent of the market.

The labs have also continued to play a role in directly commissioning, as well as buying, supercomputers. Livermore sought to commission a SOLOMON from Daniel Slotnick, only to be thwarted by lack of commitment from Slotnick's employers, the Westinghouse Corporation (Slotnick 1982; author's interview data). Then Livermore commissioned the first vector processor, the STAR 100. During the 1970s and 1980s, there was even an in-house Livermore supercomputer development. Called the S-1, this program attempted to combine sixteen pipelined vector computers, each equivalent in power to a CRAY-1, in a parallel structure. The history is fascinating. It involved the Livermore "O Group" (headed by Lowell Wood and vividly described in Broad's *Star Warriors* [1985]), the Hertz Foundation, Edward Teller, and congressional lobbying.

What effect did this heavy involvement of the nuclear weapons laboratories have on the development of supercomputing? Their need for supercomputing arises because of the difficulty of direct methods of researching nuclear explosions. These are hard to study empirically, for reasons that are simultaneously physical (the destruction of any monitoring equipment), economic (the large expense of nuclear tests), legal (tests in the atmosphere and with a yield greater than 150 kilotons are now illegal under the Partial and Threshold Test Ban Treaties) and political (they tend to be unpopular). Nor is it clear how such an extraordinary physical phenomenon as a nuclear explosion could realistically be simulated physically—though there is certainly interest in this.

Short of actual explosions, computer modeling is the most important tool of the weapons designers—"compute or shoot," as designers phrase it. Two main types of model are used: the mesh model and the Monte Carlo model.

The mesh model involves modeling the evolution of a physical quantity or set of interrelated physical quantities in a region of space. The behavior of quantities is understood to be governed by partial differential equations, but non-linearities prevent these being solved analytically. So a numerical solution is attempted by superimposing a mesh of subdivisions in the relevant space, transforming the relevant partial differential equations into finite difference equations, and calculating for a series of time-steps the changing values of the physical quantities for all the points in the mesh.

What is involved computationally can be seen by considering a typical early 1960s nuclear weapons design code (author's interview data). The mesh for this would be two-dimensional, with perhaps 50 x 50, or 2,500, cells. (Three-dimensional meshes have only been used much more recently.) Some fifteen physical quantities per cell would be computed, and the requisite calculations for one time-step for one cell might amount to 200 floating-point operations.

Such a code would require a memory size of at least 37,500 words (i.e., 2,500 times 15) to store all the values of the variables. A single time-step of the model would require half a million floating-point operations (i.e., 2,500 times 200). This puts such a model beyond reasonable implementation on a 1950s computer. The value of the IBM Stretch to nuclear weapons designers was that it made implementing such a model practically possible.

The scaling properties of this sort of mesh-computation problem meant, however, that "improvements" to the model quickly translated into greatly increased demand for computer power. Simply getting greater accuracy by using twice as fine a mesh, for example, meant a fourfold increase in memory capacity and arithmetic speed. Introducing new variables to the model, and moving to three dimensions, had even greater effects.

Mesh-computation problems most vividly illuminate the roots of the nuclear weapons laboratories' apparent insatiable demand for computer power. They are, however, problems characterized by computational predictability with relatively few data-dependent branches in the programs. Monte Carlo simulation, the other major type of computational problem of nuclear weapons design, presents quite different characteristics. Unlike the deterministic mesh model, this is a probabilistic technique. Now widely employed in physics, it was developed at Los Alamos by John von Neumann, following a suggestion by Stanislaw Ulam, for the analysis of problems such as neutron scattering and the development of a nuclear chain reaction. On account of its probabilistic nature, it was christened "Monte Carlo" by Los Alamos physicist and computer scientist Nicholas Metropolis (Hurd 1985; Metropolis 1987: 10; see also Metropolis and Ulam 1949).

High-precision Monte Carlo modeling makes heavy computational demands: three-hour supercomputer runs are common (author's interview data). The nature of the computation is quite different from that for a mesh problem: up to 30 percent of the instructions in a Monte Carlo program may be branches (author's interview data).

How has the heavy involvement of nuclear weapons laboratories affected the evolution of supercomputer architectures? I would argue that the nuclear weapons laboratories, with their heavy and permanently expanding demands for floating-point arithmetic speed, have been the prime factor in the creation and definition of supercomputing. If they had not existed, we would doubtless have a class of most powerful computer, but the criterion of power might well have been different.

A competing criterion actually existed. It too rested on the use of the supercomputer as a technology of power, for a quite different use—code-breaking. For the code breakers, most importantly the U.S. National Security Agency, the emphasis, in the words of one of their main computer specialists, "was on manipulation of large volumes of data and great flexibility and variety in non-numerical logical processes" (Snyder 1980: 66).

How the demands of the nuclear weapons laboratories came to override those of the National Security Agency, as they appear to have done, remains to be investigated. It does seem, however, that the former were the larger market. Stretch was made a cumbersome project in part by the attempt to satisfy both sets of demands, but while nuclear weapons laboratories bought four Stretch computers, only one went to a code-breaking organization. Furthermore, the demands of other likely supercomputer users, e.g., aerospace or weather forecasting, were closer to those of the nuclear weapons laboratories than to those of the National Security Agency.

Beyond this general influence in framing the meaning of "supercomputer," it is paradoxically hard to identify a specific influence of the nuclear weapons laboratories on supercomputer architecture. There are some very detailed ones, to be sure.[10] But at the level of the overall features of supercomputer architecture listed above, it is hard to trace an effect of the laboratories' particular computational tasks.

The obvious candidate is the STAR-100, the first vector computer. It was directly commissioned by Livermore, and the design of the machine was optimized for the handling of long vectors "common to many scientific problems at the Lawrence Livermore Laboratory" (Hockney and Jesshope 1988: 21).

There is a striking paradox here. The STAR-100 represents the peak of the nuclear weapons laboratories' influence on the development of supercomputer architecture. Yet the outcome was a machine perceived at Livermore and Los Alamos as ill-suited to their computational needs. Its offspring—the CYBER 205 and ETA[10]—retain its distinctive optimization for long vectors. That indeed is the most prominent performance difference between the CYBER 205 and its direct competitor the CRAY-1: "the CRAY-1 is particularly efficient on relatively short vectors, whereas the CYBER 205 is not very good at processing short vectors but exceeds the performance of the CRAY-1 when processing vectors with more than around 300 elements" (Ibbett and Topham 1989: vol. 1, pp. 189–90).[11] Yet the Laboratories eschewed the CYBER 205 and its ETA[10] descendant.

How did this paradox—the disowning by the nuclear weapons laboratories of their major legacy to supercomputer architecture—come about? The answer to this question shows that "computational needs" are neither simple nor self-evident. The STAR-100 was designed according to a particular vision of these "needs," a vision that ultimately could not be sustained.

Much of the laboratories' computational work is highly classified. Computer designers generally do not possess high security clearances and so could not have access to actual weapons design codes. Thus designers lacked immediate contact with the "needs" they were trying to satisfy. The solution for this problem attempted with the STAR-100 was to declassify 5,000 lines of the then-dominant Livermore nuclear weapons design code "Coronet." These were passed to the machine's designers. Later, however, this declassified sample was judged untypical of Livermore weapons codes. It contained too few branch instructions: in particular, it contained no data-dependent branches (author's interviews).

That made it too "easy" a test for a pipelined vector computer such as the STAR-100. The STAR's designers could satisfy the Livermore "need" expressed in the declassified code, and yet produce a machine that was only successfully run on twelve out of the several hundred Livermore weapons codes. "The STAR met our specifications, but not our expectations," was how Livermore staff put it (author's interviews).

There was nothing absolute about this "failure." Algorithms could have been redesigned, and codes rewritten, so as to make them more suitable to the architecture of the STAR-100. Indeed to a limited extent this did happen, and as the years have gone by, and vector machines become the norm, the nuclear weapons laboratories have learnt how to vectorize even seemingly quite intractable problems of Monte Carlo simulation (author's interviews). But the specific nature of the central task of the laboratories introduces a difficulty in making "the algorithm fit the architecture," just as classification causes problems in making the architecture fit the algorithm. Any weapons simulation involves approximations. Those embodied in existing weapons design codes have been "calibrated on"—their empirical validity has been checked in nuclear weapons tests (author's interviews). To change algorithms radically would involve making use of new approximations, which would, in the opinion of Livermore interviewees, require test validation.

With powerful constraints on numbers of nuclear tests, wholly new weapons design codes are thus now a rarity.[12] Here we have a specific reason for reluctance to shift to radically new computer architectures, a reason over and above the pervasive "dusty deck" problem of heavy investment in existing codes (see Johnson and Durham 1986). The laboratories' weapons design work has thus been a strong source of architectural inertia, a factor that helps to explain why the laboratories have not been in the lead in pioneering or sponsoring new computer architectures over the past decade. Evolutionary developments in mainstream supercomputing can be absorbed, but more radically parallel architectures have to date been restricted to an experimental, rather than a "work-horse," role. That may well change over the next decade, but it has not yet done so.

Durkheim and Mauss Among the Microchips

Let me end with a final speculation about the architecture of computer systems. One of the most general hypotheses of the sociology of knowledge is the claim that there is a homology between the structure of a society and how it classifies the natural world. Durkheim and Mauss (1963; first published in 1903), who first advanced the claim, noted a series of instances typified by the example of the Zuñi where "the division of the world is exactly the same as that of the clan within the pueblo" (1963: 44). These scholars believed this to be a general pattern, and that within it the structure of society was the "causal factor." There are difficulties with the argument, both analytical and in the evidence Durkheim and Mauss draw upon. Nevertheless instances have been found within scientific thought that conform to this pattern (Bloor 1982).

Tantalizing clues exist that something of this kind may be at work within computer architecture, especially when one considers more radical forms of parallelism. The central problem of these forms is often seen to be how to ensure coordination between separate processors. This bears a clear analogy to the problem of organizing society, perhaps in particular the problem of organizing human labor processes. "Social" categories—democracy, hierarchy, assembly-line production, etc.—are frequently drawn upon by participants to explain and justify particular computer architectures (see, e.g., Watson 1988). Indeed, they have a long historical pedigree—Babbage was, after all, a political economist, and student of the labor process, as well as would-be computer architect. So the speculation naturally arises: do computer designers' preferred architectures have anything to do with their preferred modes of social organization? Or, putting the matter more generally, are changes in social structure reflected in changes in computer architecture?

My research to date, I confess, suggests that the answers to these questions must predominantly be negative. But there is at least one intriguing supporting instance from the development of "radical" parallelism. It concerns the influence of that of Carl Hewitt's "actors" model of computation. By the mid-1970s, Hewitt, a member of the Artificial Intelligence Laboratory at the Massachusetts Institute of Technology, had become dissatisfied with the dominant model of artificial intelligence as akin to problem solving by an individual human being. He proposed "the problem solving model of a society of experts to supplement the model of a single very intelligent human. . . . The long term goal is to construct systems whose behavior approximates the behavior of scientific societies. That is, the ultimate aim is to build systems which model the way scientists construct, communicate, test, and modify theories" (Hewitt 1977: 324–25).

Hewitt's reasons for introducing this explicitly social metaphor remain to be researched. His own development of it, into ideas of "open systems" and into collaboration with sociologists of science, is outside the scope of this paper (though see Hewitt 1985). Interestingly, though, the metaphor was picked up by the designers at the California Institute of Technology of one of the most important classes of radically parallel machines, the hypercube. Hewitt's idea of computational "actors," communicating by message passing (Hewitt 1977), helped encourage these designers to eschew more moderate forms of parallelism, in which different processors would share a common memory, and opt instead for a system in which multiple processor-and-memory combinations passed messages from one to another. The term "hypercube" refers to their choice of topology for the network of interconnections along which these messages pass. One of those involved writes:

> We sometimes refer to the Cosmic Cube [the prototype hypercube] as a homogeneous machine. . . . C. A. R. Hoare's communicating sequential pro-

> cesses notation, the actor paradigm developed by Carl Hewitt, the processing surface experiments of Alain Martin, and the systolic algorithms described by T. H. Kung, Charles Leiserson, and Clark Thompson encouraged us to consider message passing as an explicit computational primitive. . . . Together with early simulation results, a workshop on "homogeneous machines" organized by Carl Hewitt during the summer of 1981 helped give us the confidence to start building an experimental machine (Seitz 1985: 33).

This work at the California Institute of Technology was, in turn, the inspiration of the hypercube developments by Intel, NCube and others that, currently, form the commercially most important strand of radical parallelism in the United States.

This is a difficult instance to assess, and Hewitt's "actors" model is only one among the sources cited. It is, also, only one instance. While the operation of a general "Durkheim-Mauss process" must be doubted, my suspicion nevertheless remains that the use of social metaphors in computer architecture is a topic deserving of greater attention than has to date been devoted to it.[12]

NOTES

1. In "fixed-point" representation all numbers are written with the decimal or binary point in specified position. If, as is normal in computers, the representation is limited to a given number of digits or bits, problems can arise when one program requires the accurate calculation of both very small and very large numbers. In "floating-point" representation, a number is expressed in the form

$$x = a \text{ times } 10^b$$
or
$$x = a \text{ times } 2^b$$

according to whether decimal or binary arithmetic is being used. *a* is a predetermined number of significant digits known as the mantissa, and *b* an integer known as the exponent. For example, 125,000 could be expressed in decimal floating-point representation as 1.25×10^5, or 0.00125 as 1.25×10^{-3}.

2. Measures of floating-point performance, while influential and often quoted as if unproblematic, hide a Pandora's box of issues—most obviously, what kind of floating-point operations, performed with numbers expressed to what degree of accuracy. Two participants write of "Lies, Damned Lies and Benchmarks" (Shepherd and Thompson 1988). There is a strong analogy between the issues raised in measuring a computer's performance and those raised by technological testing more generally. For the latter see MacKenzie (1989).

3. There are intriguing parallels with Bruno Latour's concept of a "center of calculation," (Latour 1987). Note, however, that supercomputers do not typically operate by summarizing and totalizing already gathered "traces," though that is obviously the role of much of computing (say of the commercial kind). They are used, as noted above, primarily for modeling or simulation. The process goes a step beyond the physically constructed models described by Latour (1987: 230–32), but the intended effect is the same. Supercomputer modeling is designed to enable the

prediction and control of the world by reproducing some of its features in a different medium—in this case an electronic medium—and to a different scale. I shall discuss below how this operates in the case of nuclear weaponry.

4. The "clock" is part of the control logic of most computers. It produces signals at a regular rate which are used to synchronize the operations of different parts of the computer. Clock speed heavily affects overall processing speed, and is itself affected both by the size of gate delays and by the length of the interconnections between various parts of the computer.

5. This "incremental/radical" dichotomy has to be used with care, because what is incremental change to one person may be radical to another. However, both the technical literature (e.g., Ibbett and Topham 1989) and participants in conversation seem to concur that the supercomputing field can be divided in roughly this way.

6. The qualification "of the corresponding period" needs emphasizing. A 1960s "radical" architecture no longer looks "radical" set against a late 1980s mainstream supercomputer with vector processing facilities and multiple central processing units.

7. ARPA, the Department of Defense's Advanced Research Projects Agency, had funded the development of ILLIAC IV.

8. This is not an "absolute technical characteristic." Thus high electron mobility transistor (HEMT) field-effect devices have been developed, primarily in Japan. Radar applications of HEMT, rather than computing applications, seem the current focus of interest, at least in the United States.

9. This argument has been interestingly put by Arthur (1984): "Very often, technologies show increasing returns to adoption—the more they are adopted the more they are improved. . . . When two or more increasing-returns technologies compete for adopters, insignificant 'chance' events may give one of the technologies an initial adoption advantage. Then more experience is gained with the technology and so it improves; it is then further adopted, and in turn it further improves. Thus the technology that by "chance" gets off to a good start may eventually 'corner the market' of potential adopters, with the other technologies gradually being shut out."

10. Such as the "noisy mode" (greater truncation) facility on Stretch, or the "normal zero" (the exponent of a floating-point number being set to zero if the mantissa is zero) on the CRAY-1 (author's interview data).

11. Hockney and Jesshope (1988: 86) make use of the concept of the "half-performance length," $n^{1/2}$, of a vector computer; $n^{1/2}$ measures how long vectors have to be before half maximum performance is achieved. They calculate a value of $n^{1/2}$ for the CYBER 205 of about 100 and for the CRAY-1 of 10–20 (*ibid.*: 98).

12. See the special issue of IEEE, *Transactions on Systems, Man, and Cybernetics,* "Natural and Social System Metaphors for Distributed Problem Solving" (IEEE 1981).

REFERENCES

Arthur, W. Brian. 1984. "Competing Technologies and Economic Prediction." *Options*, pp. 10–13.

Barnes, Barry. 1983. "Social Life as Bootstrapped Induction." *Sociology* 17: 524-45.

Barnes, George H.; Brown, Richard M.; Kato, Maso; Kuck, David J.; Slotnick, Daniel J.; and Stokes, Richard A. 1964. "The ILLIAC IV Computer." *IEEE Transactions on Computers* C-17, 8: 746–57.

Bashe, Charles J.; Johnson, Lyle R.; Palmer, John H.; and Pugh, Emerson W. 1986. *IBM's Early Computers*. Cambridge, Mass.: MIT Press.

Bloor, David. 1982. "Durkheim and Mauss Revisited: Classification and the Sociology of Knowledge." *Studies in the History and Philosophy of Science* 13: 267–97.

Broad, William J. 1985. *Star Warriors*. New York: Simon and Schuster.

Cray, Seymour. 1988. "What's All This About Gallium Arsenide." Videotape recording.

Donaghue, Hugh. 1987. "A Business Perspective on Export Controls." In *Selling the Rope to Hang Capitalism? The Debate on West-East Trade & Technology Transfer*, ed. Charles M. Perry and Robert L. Pfaltzgraff, Jr. Washington, D.C.: Pergamon-Brassey.

Dosi, Giovanni. 1982. "Technological Paradigms and Technological Trajectories: A Suggested Interpretation of the Determinants of Technical Change." *Research Policy* 11: 147–62.

Durkheim, Emile, and Mauss, Marcel. 1963. *Primitive Classification*. London: Cohen and West; first published in French in 1903.

Falk, Howard. October 1976. "Reaching for a Gigaflop." *IEEE Spectrum* 13, 10: 65–69.

Flynn, Michael J. 1972. "Some Computer Organizations and their Effectiveness." *IEEE Transactions on Computers* C-21, 9: 948–60.

Hewitt, Carl. 1977. "Viewing Control Structures as Patterns of Passing Messages." *Artificial Intelligence* 8: 323–64.

————. 1985. "The Challenge of Open Systems." *Byte* (April).

Hockney, R. W., and Jesshope, C. R. 1988. *Parallel Computers 2: Architectures, Programming and Algorithms*. Bristol: Adam Hilger.

Hughes, Thomas P. 1983. *Networks of Power: Electrification in Western Society, 1880–1930*. Baltimore: Johns Hopkins University Press.

Hurd, Cuthbert C. April 1985. "A Note on Early Monte Carlo Computations and Scientific Meetings." *Annals of the History of Computing* 7, 2: 141–55.

Institute of Electrical and Electronics Engineers [IEEE]. 1981. *IEEE Transactions on Systems, Man, and Cybernetics* SMC-11, 1. (Issue on "Natural and Social System Metaphors for Distributed Problem Solving.")

Ibbett, R. N., and Topham, N. P. 1989. *Architecture of High Performance Computers*. Basingstoke: Macmillan.

Intel Scientific Computers. 1989. "Intel, DARPA to Develop Prototype of 2000-Processor System Using New Intel Superchip." Swindon, U.K.: Intel Corporation.

Johnson, Tim, and Durham, Tony. 1986. *Parallel Processing: The Challenge of New Computer Architectures*. London: Ovum.

Latour, Bruno. 1987. *Science in Action: How to Follow Scientists and Engineers through Society*. Milton Keynes, Bucks.: Open University Press.

Merton, Robert K. 1948. "The Self-Fulfilling Prophecy." *Antioch Review* 8: 193–210; reprinted in Merton, *Social Theory and Social Structure* (New York: Free Press, 1968).

Metropolis, N. 1987. "The Los Alamos Experience, 1943–1954." Paper presented to the ACM Conference on History of Scientific and Numeric Computation. Los Alamos, New Mexico: Los Alamos National Laboratory, LA-UR-87-1353.

Metropolis, N., and Ulam, S. 1949. "The Monte Carlo Method." *Journal of the American Statistical Association* 44, 247 (September): 335–401.

Molina, Alfonso. 1989. *Transputers and Transputer-Based Parallel Computers: Sociotechnical Constituencies and the Build-Up of British-European Capabilities in Information Technologies*. Edinburgh: Programme in Information and Communication Technologies.

Nelson, Richard R., and Winter, Sidney G. 1982. *An Evolutionary Theory of Economic Change*. Cambridge, Mass.: Harvard University Press.

Noyce, Robert N. 1980. "Microelectronics." In *The Microelectronics Revolution*, ed. Tom Forester. Oxford: Blackwell; first published in *Scientific American* 237, 3 (September 1977).

Nuclear Weapons Databook Project. 1988. "Nuclear Notebook: Known Nuclear Tests by Year, 1945 to December 31, 1987." *Bulletin of the Atomic Scientists* 44, 2: 56.

Pavitt, Keith. 1984. "Sectoral Patterns of Technical Change: Towards a Taxonomy and a Theory." *Research Policy* 13: 342–74.

Pease, Marshall C. 1977. "The Indirect Binary n-Cube Microprocessor Array." *IEEE Transactions on Computers* C-26, 5: 458–73.

Seitz, Charles L. 1985. "The Cosmic Cube." *Communications of the ACM* 28, 1: 22–33.

Shankar, Ravi, and Fernandez, Eduardo B. 1989. *VLSI and Computer Architecture*. San Diego: Academic Press.

Shepherd, Roger, and Thompson, Peter. 1988. "Lies, Damned Lies and Benchmarks." Inmos Technical Note 27. Bristol: Inmos.

Slotnick, D. L. 1971. "The Fastest Computer." *Scientific American* 224, 2: 76–87.

————. 1982. "The Conception and Development of Parallel Processors—A Personal Memoir." *Annals of the History of Computing* 4, 1: 20–30.

Slotnick, D. L.; Borck, C. W.; and McReynolds, R.C. 1963. "The SOLOMON Computer." In *Proceedings of the 1962 Fall Joint Computer Conference*, pp. 97–107.

Snyder, Samuel S. 1980. "Computer Advances Pioneered by Cryptological Organizations." *Annals of the History of Computing* 2, 1: 60–70.

Stone, H. S. 1975. *Introduction to Computer Architecture*. Chicago: Science Research Associates.

————. 1987. *High-Performance Computer Architecture*. Reading, Mass.: Addison-Wesley.

Thornton, J. E. 1970. *Design of a Computer: The Control Data 6600*. Glenview, Illinois: Scott, Foresman.

Watson, I. 1988. "Language First Philosophy." In Institution of Electrical Engineers, Electronics Division, *The Design and Application of Parallel Digital Processors*, pp. 83–86 (London: Institution of Electrical Engineers, Conference Publication No. 298).

REFLECTIONS AND TRIANGULATION

Three Commentaries

A View from Economic History—Louis Galambos

(Department of Economics, Johns Hopkins University)

I respond to these sessions and the whole program as one trained in history, whose work has been almost entirely on U.S. economic institutions in the period from 1880 to the present. Most of these institutions either were, or aspired to be, large-scale. Some of them failed. Most of these institutions involved a distinctive, and for their time relatively modern, technology.

To prepare for this, I reread the Cologne papers (Mayntz and Hughes 1988) and came away from that experience with the following conclusions. The LTSs being studied were for the most part what economists and historians had studied as SOC—social overhead capital. They had large capital needs and the investments were lumpy—that is, it was hard to make them incrementally and get any results. The gestation periods were long; the externalities were very important. Often the externalities outweighed the profits to be made from a system. Most were subject to economies of scale and system, and hence had declining cost curves, long-run cost curves, and were so-called natural monopolies.

The LTSs were at an early period in their development subject to substantial public concern. The result was either public operation, which tended to be equity-oriented and less entrepreneurial or efficiency-oriented than private management, or public regulation, which was a compromise between an emphasis on equity and an emphasis on efficiency and innovation. The LTSs frequently followed a path or curve of development, something like the S curve or product cycle with which we are all familiar. Their technologies experienced an early phase of experimentation followed by rapid development and extension, and, finally, by a phase of leveling off as the technology was perceived as mature (whether or not it was). New technologies could grow off of these systems and frequently they faced competition from new technologies that developed outside of their formal boundaries. These are my major conclusions from the work that we did up to that time. What, then, has been added by this conference to this general picture of the LTSs?

First, there is a new emphasis on the human element in LTS evaluation: we have in effect humanized the study of large technical systems. Frequently, these systems have been studied as a phenomenon that can or should only be analyzed in behavioral terms which focus on the forces of history. Usually these forces are technological. But they are also economic, they are political, and sometimes we use class lines of analysis. Eventually they are aggregate forces that shape what individuals do. Following this line of analysis, if we understand these forces we will understand what the people did. This style of analysis is the normal approach to economics and we frequently encounter it in economic history and in the sociology of organizations. I know that Tom Hughes has

T. R. La Porte (ed.), Social Responses to Large Technical Systems, 177–188.

always included entrepreneurship and managerial leadership in his model of the LTS. But I think there has in general been a tendency to study these systems in terms of the manner in which they were shaped by certain forces in history.

This conference, I think, has done something to advance the idea that people are involved in these developments and that frequently these individuals have played very important roles. At times that theme arose as a result of deconstruction to the individual level, where decisions were made by people about particular phenomena. At times we were presented with analyses that stressed contingency; we thus encountered people who expressed an interest in personal goals and wanted to achieve certain objectives. The human element became involved at times as a result of focusing on managerial talent in, say, energy development, so that the performance of particular people, experts who actually had names and could be identified, could be evaluated. They said stupid things at times. We could actually see these people making decisions. The personal element also entered into our study of public policy choices and, most memorably, in the analysis of unique human functions in the operation of dangerous LTSs in the Navy and in air control. The bubble idea that was so captivating to most of us involved the reintroduction of a very important human phenomenon into the setting of the LTS. If this aspect of our conference can be considered a trend, this trend fits with what I see happening in organizational history in general.

Scholars have mapped patterns of institutional development and the forces shaping them. Leaders have important elements of discretionary power. One of the things we have created is systems in which leaders in particular areas have very important elements of discretionary power. Whether or not to do something devastating to the economy of Syracuse by moving to someplace in Alabama may be a decision made by one man, a corporate chief executive officer. In the case of changing the largest telecommunications system in the world, the U.S. system, two men, virtually independent of any kind of control, decided what should be done. They also decided how to implement their decisions to break up the Bell System.

So there are important elements of discretionary power in the systems that we have been studying, and I think they make leadership crucially important. Moreover, organizational cultures, as well as hardware, appear to shape organizational performance. Thus we are forced to acknowledge that social systems consist of people who have to be motivated to perform effectively over the long run. The NASA experience in its different phases provides, I think, an excellent example of this—how an organization can perform well under certain circumstances and not so well under others.

There are of course different ways to look at this cultural element or this human element in LTS development. I have been studying it lately in terms of the way particular leaders (in the case of the Bell System, Theodore Vail) shaped a culture deliberately to induce technological innovation over the long run. You can see this kind of cultural and leadership function as moving *past* the Beniger control model or, as I prefer, as building *on* it. I see the behavioral model and a humanistic model fitting together as we map out how these institutions develop. We describe and analyze the forces that shape LTS development, but then we must incorporate the people in them who make choices.

The second thing that I have learned in this set of conferences is a new appreciation for the various modes of political involvement with the LTS. When I started my own study of this problem, many years ago, I had two sets of generalizations or contexts available for analyzing that political element. One was what I will call (at the risk of making some people mad) a liberal or Left context that has been extremely important in all of our analyses of government for one hundred to one hundred fifty years, both in shaping government and in analyzing it, and in evaluating and explaining what should be done and what was being done to and for LTSs. What this viewpoint tells us is that it is good to have public ownership or control, so that externalities can be handled and equity concerns introduced in the operation and evolution of the LTS.

In American history this has been the dominant viewpoint. In this model, there are bad actors like big telecommunications systems or railroads, and they come along and crunch the little people; the little people protest and then they form organizations, and eventually they create a reform movement. As a result, everybody is better off and history ends at the New Deal—there was Truman and Eisenhower, but the important history really ends with the New Deal and its measures of liberal reform. That has been a very popular view of history. History comes in cycles as it gradually gets better and better. So it slopes up. More recently, another context has been developed for looking at the same phenomena. It is a neoclassical or "Chicago" context that has focused attention on the inefficiencies these liberal developments have introduced; in particular, this new model focuses attention on the inability of regulated or public systems to innovate. After the initial surge of LTS development, the regulated or public systems seem to have had a great deal of trouble responding to new technologies. Often governments united two technologies within the same regulatory system. The telegraph and telecommunications example is a very good one in this case; equally good in the United States is that of the railroad and trucking systems, which were brought under the same regulatory umbrella to bring the new system under control.

What am I to make of those two contexts when I reflect on the results of this conference? Clearly, many of the LTS scenarios that were developed here have episodes that fit one or both of these contexts. They also have aspects that do not fit either one of them terribly well. So, I am left with some problems. I would like to see more consideration of the long-run ability to sustain innovation within an LTS and more thought given this question of innovation from outside the LTS. The information in the conference suggests to me that most large-scale public or private bureaucracies have had trouble innovating over the long run, seeming thus to support the neoclassical or Chicago school analysis. The systems have a tendency to ride the S curve, to stay near that S curve. But of course not all do and we have had examples in the conference of those that do not. Then we have a problem: how are we going to analyze the behavior of big private and public bureaucracies that do not follow this pattern? We need to understand why they do not. There is work now being done on corporate reorientation and revival, on how IBM and National Cash Register made the transition from keypunching and mechanical equipment into electronic equipment, when Addressograph did not. These kinds of comparative studies of large institutions are being conducted, and I think that kind of analysis needs to be pushed here. What the program produced for me was many subtle patterns that did not fit terribly well in either one of the two rather gross contexts I have described. Andersen's work on Norwegian corporatist innovations was very interesting in that regard; McKenzie's work on innovation and supercomputing was also very interesting, as was Bijker's work on fluorescent lighting (which I saw in a slightly different way than he did, while his work nonetheless prompted me to make one of my big concessions to socially constructed technology). Salsbury's work on the WPPSS, Schneider's and Videll's on telecommunications, all suggest to me that the gross patterns need to be changed. We need models that allow for more variability in public sector behavior and in the behavior of large-scale private bureaucracies. We need to acknowledge and understand public sector innovation as well as the efficiencies of certain forms of public control and organization.

Finally, I concluded from this conference that one of the things we most need is a vocabulary for dealing with culture and values. I confess that I still use structural-functional categories. I have a heavy investment in Talcott Parsons, and I even wrote a translation once of Talcott Parsons into English. I still use structural-functional categories (but not necessarily the model because it has no dynamic properties). I find the categories quite useful. In this conference I found all of us looking for ways to compare cultural components that we now realize, as we develop these histories, shape LTS development and operation; they also shape the public responses to them, but we find this hard to analyze because we do not have a very good vocabulary for dealing with culture. This came out in all the cross-national comparisons and in our very pointed discussion of what Americans would and would not do. The explanation turned on questions of culture. Technologies, we find, are shaped by cultures as are the systems used to control them; at least they are influenced by them. But we find that part of the problem much harder to talk about and analyze than we do either hardware or formal organizations. I hope we can develop better conceptual tools for analyzing culture so that we can achieve a richer, more balanced view of LTS evolution.

Discussion

The discussion centered on the degree of adequacy of the existing economic models for describing LTSs. There was agreement that neither the liberal/Left division nor the Chicago-school traditions were satisfactory. It was argued that they "put LTSs in nice little categories which are not very helpful in describing people's thinking about what is going on." For example, private companies and the government bureaucracies often have many properties in common and it's sometimes very difficult to tell the difference between their various behaviors. The models we have are sufficiently crude that they have a very difficult time accommodating what Tom Hughes approvingly calls "the complexity of history and the particularism of the LTSs." Yet we do need models to synthesize and analyze the "incredibly complex phenomena of LTS developments."

The discussion then turned to redressing the inadequacies of contemporary models. The conference generally applauded Galambos's emphasis on the importance of the human element, contrasting its relative absence with perhaps too much emphasis on the supply side, too little attention to the users. The major users—and those who oppose development—have to be integrated into the modeling process. Only in dramatic cases, like nuclear power, do we begin to look at these actors a little bit more. Given the data we have, it's a lot easier to concentrate on the supply side.

Gene Rochlin introduced a concern about the implications of our discussion on theory development. He wondered if the characteristics of LTSs represent a change in the natures of their various systems and societies from what they were in the days of the classical modelers and users. Because of the technological changes that have occurred in the background, the users have gained access to power. They've gained access to each other. He argued that in the old models, the users were a dispersed community of unlinked actors, each one responding to a central stimulus. If this is not built into the old models, they're no longer very good. The alternative is probably not between the classical models or the constructionists entirely; yet there is no viable alternative and in attempting to develop theory that takes into account a new reciprocity and a sort of empowerment, there is the risk that the whole middle range of models that have not been developed and are not available would be cast into doubt. This is true in analyzing energy and environmental issues: the models are not there that take into account the new empowerment of user communities—something that should catch our attention.

Galambos agreed and nominated two areas of work that are relevant to the problem: the interest group approach to government, and the growth of the state in America. In the first area, the approach is to analyze the tendency of populations to break down into functional interest groups and for these groups to shift the nature of the politics. The usual assumption is that change is incremental and everything works out for the best—an assumption which the social deconstructionists could work over. The other area (the growth of the state in America) includes one topic that was raised a number of times—the politics of administration and the politics of staff activities. These politics challenge the current view of how the state is run, regarding especially the dynamics of congressional staff in decision-making in the U.S. Congress. Staff contacts and staff lobbying are at the heart of the new kind of politics—dynamics that can be woven into the study of LTSs because of the nature of the systems and the political controls that accompany them.

A View from the Social Sciences—Renate Mayntz

(Max Planck Institute for Social Research, Cologne)

My assignment involves not so much the search for new topics, but a critical assessment of the presentations and discussions over the past days. This, of course, does not mean an attempt to summarize what we all have heard, but rather to point to the specific challenges to research and theory-building which in my view have emerged in these discussions. We have tried to draw, these past days, a very sketchy map of a new continent which needs considerable refinement before we can set sail for new shores. The biggest challenge to future research and theory-building in this field is therefore not to switch to new topics, but to continue developing the ones we have already touched upon in a more systematic way. What we are lacking is above all systematic comparative research on some of the basic issues we have been dealing with.

Our material is presently mainly of the case-study type, and the conclusions are sometimes highly suggestive, but I keep asking myself whether they would also hold true for a different type of technical system, or for that same type of system in a different country, or fifty years ago. There are deficiencies on all these dimensions for comparison—across countries, across technological systems, and along the time axis in the development of given technical systems.

Most of the issues which I have put down in reviewing our discussions so far refer to what has been called the supply side, neglecting the users as it were. However, I think that is not a decisive deficit because users are in fact an integral part of the systems we were talking about; without users, these systems would not exist. Nevertheless, there is indeed a certain bias in our discussions of LTSs to focus explicitly on the systems themselves rather than on their users.

Over the past days, we have been talking mostly about structural issues of LTSs. In particular, we need more systematic knowledge about control structures of large technical systems (no matter how "control structure" may be defined at the moment). The control structures of different technical systems can, of course, differ from each other, so control structure should be explicitly considered a variable.

Talking about LTS control structures means, more specifically, to ask about the relationship between *structure* and *performance*, i.e., the control capacity of different types of structure. Though in the papers we discussed the control performance of a given system has now and then been linked explicitly to specific structural features, these interrelationships have not been treated systematically in a comparative fashion. To do so would, of course, require a more differentiated treatment of the two major variables involved.

To begin with structure, some typological efforts have already been made. This morning, for instance, a distinction has been made between hierarchy, two-dimensional, and three-dimensional networks. In discussing LTS structure, we may have been inclined to focus too much on types of hierarchical or centralized control, and have tended to frame the issue in terms of centralization/decentralization. I think we should include explicitly different modes of self-regulation of technical systems, though that might be something not easily found among hard-wired LTSs.

It seems also necessary to make more systematic distinctions on the performance side, the dependent variable so to speak. Here, however, it is important to realize that in assessing system performance, we are really interested in more than one effect or outcome. Thus, we can ask whether a system is failure-proof or not, risk-avoiding or not, but we are also interested in its effectiveness in fulfilling its official function, e.g., the supply of a service, and in its economic efficiency. We might even be interested in the responsiveness of LTSs to changes in demand. In trying to make links between control structures and performance, we should therefore systematically determine what kind of output variable we are mainly interested in, because it is quite likely that different structures will be optimal for different kinds of performance aspects. This points to the likelihood of conflicts in system design, e.g., whether to maximize effectiveness or safety. The fact of such conflicts is well known, but I do not see that we have pursued this relationship in a systematic way.

Under the very general heading of the structure-performance relationship there is a number of more specific issues which may serve to illustrate the more specific issues of the research challenge that I see. One of these is the design issue, i.e., the question of how the anticipation of consequences along the different performance dimensions I have mentioned determine system design. What do system designers really anticipate? Which of the various consequences do they anticipate? And if they anticipate them, how does that enter into the design process? The system builders, of course, are not only the engineers. They include the politicians who regulate a system before it ever goes into operation. To try to answer such questions systematically by comparing a variety of cases would be extremely challenging, but theoretically helpful.

That also applies to another question raised both in this morning's discussion and yesterday: What changes does the use of modern information technology introduce into the control processes relating to different aspects of system operation and performance-effectiveness, risk of failure, and negative externalities? In this connection, we have heard some highly interesting hypotheses, for instance Gene Rochlin's thesis that the introduction of modern information technology leads to a redistribution of control functions between technology and human beings. He has argued more generally that the availability of sophisticated monitoring equipment may tempt military leaders to seek full hierarchical control instead of making do with management. Modern information technology is thus seen as a force counteracting a trend toward the decentralization and self-regulation of technical systems. Gene's point was made about the military and in a historical perspective, but would it be true just for the military? Is it also true in other technical systems? Would even the military agree that it is true everywhere or not? Again, these questions pose the challenge of trying to collect systematic evidence on the case-based hypotheses offered. Or to take one more example, we were discussing this morning to what extent the use of modern information-processing technology may obviate the need for a (relatively crude) standardization of informational inputs for specific control processes. Does that in fact happen, and if so, where does it really happen and what is this effect contingent upon? Again, you can see how much work it would take to come up with an empirically founded generalization in response to these questions.

I want to close by briefly raising an issue that is more a topic for a theoretical debate which I myself would be very interested in, a debate linked to the tension between the different approaches represented around this table. I would formulate this particular theoretical challenge in terms of the question: How can the reconstruction of micro-events, with due attention to all the processes of interpretation that go on at this level, be connected with macro-structural analyses? How does the former feed into the latter? The tension between the different approaches we are using is not necessarily solved best by saying that we are asking different questions (and are therefore using different approaches), or that we are asking questions on different analytical levels. I think we might just as well ask whether—and if so, how—the two can be fruitfully connected. I am personally convinced that they can be connected, but I realize that we have not thought this through very systematically. This, then, is in fact more a challenge for theory-building than for research.

Discussion

In the discussion, the last point was taken up first. The question was raised whether what is at issue is really the difference between micro and macro; in fact, these terms would need to be defined to carry the argument forward. Thus, constructivist analyses may well involve long periods of time, several thousand people, and several different countries. The issue might therefore be rephrased to ask how a detailed analysis of technical change or development feeds into the analysis of macro-events such as the success or failure of large-scale projects.

With respect to the main point, the need for more systematic knowledge about the relationship between control structure and control performance, Gene Rochlin briefly sketched the analytical scheme for describing control structures that he presented in the final discussion on the last day. Todd La Porte emphasized the importance of such analytical work by pointing out that we particularly lack a conceptual language for the adequate description of (socio-technical) *networks*. We try to describe networks in terms of the social patterns that are replicated within a large technical system, we can picture complex networks in our minds, but when we discover that networks are not only horizontal but vertical in their messy lines of interaction between people and groups, with different degrees—or different intensities—of coupling, we are lacking adequate categories for description. Distinctions such as the one between hierarchy and market are insufficient for capturing the intricacies of network structure.

The remainder of the discussion was sparked by a protest against the commentator's emphasis on the need for *systematic* knowledge. Quite in contrast to the commentator's emphasis on the need for systematic research, Trevor Pinch asserted his option "to go for things like wild ideas, to be more eclectic." Systematization involves not only gains, but also losses. Thesis and antithesis were followed by synthesis, i.e. the assertion that systematic and event-centered approaches complement each other. In a less confident vain, Bernard Joerges argued that while "it may not be easy to be very systematic about very messy things," we should at least be careful in making distinctions and in attributing a specific meaning to the words we employ. Referring to an earlier statement that there can be no intrinsic distinction between technology and politics, because there exists no bedrock of solid technical fact, Joerges argued that this could be maintained only at the level of talk about technology and talk about politics, but not with respect to the technical object, the real artifact. This insistence on a bedrock of solid (technical) fact was challenged by the constructivist argument that all facts are interpretations, and that this holds particularly for technical systems which are *intended* for some conceived purpose, so that perceptions of events (like a missile hitting the ground) will always be infused with elements of evaluation and the attribution of purpose. For that reason, it is manifestly difficult—though we often do not notice it—to make neutral physical statements about any technology. Statements of this type that can be made, e.g., that the temperature of a given liquid has been increased by two degrees, are sociologically uninteresting. Though artifacts are physical objects subject to the laws of science, that is not the interesting thing about them to us, and it is the constructivist approach that reminds us of this.

Historical Overview—Thomas Hughes

(Department of the History and Sociology of Science and Technology, University of Pennsylvania)

I have three objectives in these brief remarks. The first is to remind us of the overarching themes that have structured our discussions during our conferences in Berlin, Cologne, and now Berkeley. The second is to suggest the significance of the study of large technical systems. In this connection, I shall call attention to several influential books that have explored systems and society in some depth. Lastly, I shall suggest that we are using metaphors and vocabularies strange to many of our social science and historian colleagues as we explore systems and society. These remarks, I fear, will be all too superficial; we need Lou Galambos to do one of his overview, state-of-the-art articles for us, essays for which he is justly celebrated.

In 1985, before our first Berlin conference, Todd La Porte and I discussed the objectives of the large technical systems conferences we envisaged. In this endeavor we were staunchly supported by George Thurn, research director at the Wissenschaftszentrum in Berlin. After these discussions Todd wrote a prospectus calling for a biennial conference series on the evolution, dynamics, and policy implications of LTSs. In this document (summarized in part in the introductory chapter), he stressed that the subject was of the utmost importance because the number, scale, scope, and complexity of the systems have increased steadily over the past century. We found scholars from various countries and disciplines who agreed that the study of LTSs was of the utmost importance, and our conference series was launched. I suggest that this 1985 prospectus be placed in our archives. Another candidate for our archives is an analysis I did of our Berlin discussions. From these discussions emerged thematic clusters that still pattern our deliberations—dynamics of systems growth, management and control of systems, and the comparative study of systems. Before our next conference in Australia in July 1991, these documents will be circulated so that we can begin to speak of our intellectual constitution—subject, I hasten to add, to revisions and amendments.

Now I shall turn to my second subject, the significance of our field of study as seen by several scholars and as developed in their influential books. I would like first to make a few general remarks on the pervasiveness of modern technological systems. They are among the deep structures of the modern world. These material constitutions of our world may now be more influential than the political ones; Jacques Ellul in his various works has also argued this. Technological systems surely shape history in the industrial world more than geographical (natural) features. Technological structures are now analogous to the geographical ones that Vasillii Kluchevskii, a historian of Russia, so eloquently described in the first chapter of his monumental history, for technology now provides the stage for the human drama like Kluchevskii believed geography did in previous centuries. Modern historians should open their books with descriptions of the technological frame.

Systems of energy, communication, waste disposal, and transportation are omnipresent. They help determine where we live, work, and play. They deeply shape our social systems. The layout of our cities, factories, and houses depends on the systems. War is made using them. We could elaborate on this theme at great length. Nature was once the stage of history; technology becomes increasingly so. The point is made in a book by Bill McKibben, *The End of Nature* (1989).

Rosalind Williams, a professor at Massachusetts Institute of Technology, published *Notes on the Underground* in 1990, a book about the world as a human-made artifact that I expect to be influential. She develops an extremely imaginative concept as she argues that the constructed underground is the essence of the non-nature of the technological creation. She writes about the undergrounds—underground (subway) systems—that we have made with our technology and the undergrounds that authors have imagined in their novels. She demonstrates that many authors have placed utopias and dystopias underground and, in so doing, have isolated them from the given, or natural, world. The underground is a fine metaphor for the purely technological world.

Turning to earlier works about the systems of our human-made world, Andrew Ure, writing in around 1830 about the factory as a system, conceived of it as an automaton. To him, the workers in a factory were only an extension of the machine. The factory for him was a human-mechanical system. He wrote explicitly of systems. Karl Marx in *Capital* further developed this idea and argued that the means of production should be conceived of as a system, as a machine. He seems to have been fascinated by the interdependence and interaction of components in this system. Perhaps we shall have a paper at one of our conferences on Marx's thoughts on technological systems.

In *Capitalism and Material Life* (1973), Fernand Braudel, a twentieth-century French historian, also explored the systems that constitute the material world. Earlier, in *The Mediterranean and the Mediterranean World* (1972), he had looked for the geographical structure. So his interest shifted from the natural to the technological stage of history. Lynn White, in his *Medieval Technology and Social Change* (1962), presented medieval manorial economy as a system. He argued that the introduction of the deep plow resulted in an improved harness, and that together these necessitated the multi-animal plow team. These interacting components accompanied the change to a new social organization—the institution of the communal manor. When I told Lynn that he was a historian of systems, he replied that his school of management would be delighted to learn this.

Walther Rathenau, a twentieth-century German industrialist, banker, social critic, and technological system builder, also saw technology as systematizing the world. In *Zur*

Mechanik des Geistes [*The Mechanization of the Soul*] (1913), he explored the world of mechanization, or systematization, and the realm of the soul, or the spirit. This is a dichotomy explored by other modern authors concerned about the mechanization of society. Some refer to a similar conflict that they see between the mechanistic and the organic. It has not, however, been a theme raised in this conference. Rathenau did not use "mechanization" in the way Marx and Ure employed it. Rathenau used the work "mechanization" to refer to the flow of electricity, water, roads, railroads, and other energy and communication networks into and out of the city. So in fact mechanization was for him the creation of large technical systems. Thorstein Veblen, like Rathenau, also believed that the world was being systematized. In *Engineer and the Price System* (1921), he envisioned the entire national economy organized into a smoothly functioning system of interrelated parts. He also called for a soviet of engineers to preside over the system because financial capitalists were unable to appreciate the interdependence of components, input/output relationships, feedback, and other production system characteristics. Martin Heidegger, a modern German philosopher, in *The Question Concerning Technology and Other Essays* (1977), is also among those who saw technology as structuring the modern world systematically. He had the deeply disturbing insight that there was a relationship between the carrying out of the Holocaust and modern man's systematizing and bureaucratizing tendency. In contrast to Heidegger, Rathenau, and Veblen, Siegfried Giedion in *Mechanization Takes Command* (1948) took a conventional view of mechanization and wrongly insisted that the twentieth-century world is characterized by machines, not systems. He, like so many other unobservant persons, believed that the machine, not the technological system, is the twentieth-century icon.

Lewis Mumford has also given us insights into the LTS. His appellation for such a system is "megamachine," which he defines as a system of both human and inorganic parts dedicated to an orderly and controlled movement to a clearly defined goal. For him, the first megamachine was the system that built the pyramids for the pharaohs. Mumford sees the various projects of the military-university-industrial complex as megamachines, hence his use of the title *Pentagon of Power* (1970). The modern megamachines contribute, he believes, to the momentum of technological determinism, and embody reprehensible values.

In our search for our predecessors who also focused on the role of large technical systems in history, I turn now to the counterculture of the late 1960s and early 1970s. In writing *American Genesis* (1989), I discovered that the most substantial thinkers among the counterculture elite identified LTSs as a salient negative aspect of contemporary culture. Herbert Marcuse, Paul Goodman, and Theodore Roszak are cases in point. Oppressive technological—not political—systems were their major concern. Out of the counterculture came postmodern thinking as well, especially in architecture. In part, postmodern architecture is a response to the order and control of large-scale modern systems. Postmodern architects, such as Robert Venturi, are stressing counter-themes to those of order and control, themes such as messy vitality, and complexity and contradiction.

Finally, I wish to suggest that those of us who study large technical systems should realize that we are using metaphors and vocabularies not familiar to many of our academic colleagues, and we are therefore likely to be misunderstood. Recently, after several conferences, I discovered—too late to correct the misunderstanding—that most of my historian colleagues assumed that the system builders of whom I spoke were and are little more than business entrepreneurs and that their system-building differed and differs little from the founding and development of business firms. It also dawned on me during discussion that my colleagues use mechanical metaphors to explain relationships while I use ones borrowed from electrical engineering. My metaphors tend to circuits, fields, and systems; theirs to mechanical trains of cause and effect. I argued that Walther Rathenau, who presided over Allemeine Elektrizitäts-Gesellschaft and the electrical utilities it owned, had a different mindset and used different entrepreneurial tactics, managerial tactics, and strategies from a business entrepreneur like Alfred Krupp, a steel magnate. I am sure that my point was taken. This episode reminds us that we may often talk past one another. We need to be constantly aware that there is a language problem, an imaging problem, a metaphor problem that we must meet if we are to talk effectively to other audiences about large technical systems.

Editor's Afterword

Tom Hughes's remarks were received with a kind of gratified silence. Discussion was anti-climactic. He had provided a fitting end to a time of vigorous and fruitful exchange.

REFERENCES

Braudel, F. 1972. *The Mediterranean and the Mediterranean World*. New York: Harper & Row.

————. 1973. *Capitalism and Material Life*. New York: Harper & Row.

Giedion, S. 1969. *Mechanization Takes Command*. New York: W. W. Norton.

Heidegger, M. 1977. *The Question Concerning Technology and Other Essays*. New York: Garland Publishers.

Hughes, T. P. 1989. *American Genesis*. New York: Viking Penguin.

Mayntz, Renate, and Hughes, Thomas, eds. 1988. *The Development of Large Technical Systems*. Frankfurt: Campus Verlag; Boulder, Colo.: Westview Press.

McKibben, W. 1989. *The End of Nature*. New York: Random House.

Mumford, L. 1970. *Pentagon of Power*. New York: Harcourt Brace Jovanovich.

Rathenau, W. 1918. *Zur Mechanik des Geistes*. Berlin: S. Fischer.

White, L. 1962. *Medieval Technology and Social Change*. Oxford: Clarendon Press.

Williams, R. 1990. *Notes on the Underground*. Cambridge, Mass.: MIT Press.

Veblen, T. 1921. *Engineers and the Price System*. New York: Viking Press.

APPENDIX

Conference Agenda and List of Participants

I. Macro Perspectives

Technologies of Scale, Variations in Social Response and Democratic Political Thought
Langdon Winner, Department of Science and Technology Studies, Rensselaer Polytechnic Institute, Troy, NY, USA

The Governance of Large Technical Systems: The Case of Telecommunications
Volker Schneider, Max Planck Institute for Social Research, Cologne, FRG

Large Scale Technologies, Risk Assessment and the Legitimacy Trap
Peter Weingart, Center for Science Studies, Bielefeld University, FRG

[Discussion of Earthquake Responses]

II. Responses to Non-Market Systems

Political Responses to Large Scale Development: Norway and North Sea Oil
Svein Andersen, Stiftelesen Norwegian School of Management, Sandvika, Norway

Technical Systems and the Military: Four Faces of Control
Donald MacKenzie, Department of Sociology, University of Edinburgh, UK

Computers and the Illusion of Control in the Military
Gene Rochlin, Institute of Governmental Studies and Energy and Resources Group, University of California, Berkeley, USA
Discussant, Arne Kaiser, Department of Technology and Social Change, University of Linkoping, Sweden

III. Scale, Hazard and the Dilemmas of "Fail-safe" Systems

Facing the Collapse of the Washington Public Power Supply System
Stephen Salsbury, Department of Economic History, University of Sidney, New South Wales, Australia

Negotiating the Social and the Technical in Systems Failure: The Case of the Space Shuttle Challenger
Trevor Pinch, Institute for Research in the Social Sciences, York University, York, UK
Discussant: Todd R. La Porte, Department of Political Science, University of California, Berkeley, USA

T. R. La Porte (ed.), Social Responses to Large Technical Systems, 189–190.

IV. Taking Control.

Technology and the Increasing Wish to Control

James Beniger, School of Communications, University of Southern California, Los Angeles, USA

The Task Force Approach to the Analysis of Major Technological Policy: The Presidential Commission on Aviation Safety

Clinton Oster, Environmental and Public Policy, Indiana University

Technology Assessment and the Dutch Chemical Industry: Contributions of the "Social Constructionist"

Wieber Bijker, Science and Technology Studies, University of Limburg, Maastricht, The Netherlands

Discussant: John Andelin, Assistant Director, Office of Technology Assessment, U.S. Congress.

Anticipating and Regulating Potential Negative Externalities of LTS: The Case of New Telecommunication Services in France, Germany and UK

Thierry Vedel, Centre d'étude de la via politique française contemporaine (CEVIPOF), Paris

V. Recapitulation and Ventilation of Neglected Issues

Responses from the Economic Historical View

Louis Galombos, Department of Economics, Johns Hopkins University, Baltimore, MD USA

Challenges to Research and Theory

Renate Mayntz, Max Planck Institute for Social Research, Cologne, FRG

Substantive Overviews of the LTS Series

Thomas P. Hughes, Department of the History and Sociology of Science and Technology, University of Pennsylvania, Philadelphia, PA USA

Other Participants

Ted Bradshaw, Institute of Governmental Studies, University of California, Berkeley, USA

Glenn Bugus, Department of the History of Technology, California Institute of Technology, USA

Bernward Joerges, Wissenschaftszentrum Berlin, FRG

Svante Lindqvist, Department of History of Science and Technology, Royal Institute of Technology, Stockholm, Sweden

Jane Summerton, Department of Technology and Social Change, University of Linkoping, Sweden